NATURE'S IMAGINATION

NATURE'S IMAGINATION

The frontiers of scientific vision

◆

Edited by
JOHN CORNWELL

Introduction by
FREEMAN DYSON

Oxford · New York · Melbourne
OXFORD UNIVERSITY PRESS
1995

Oxford University Press, Walton Street, Oxford OX2 6DP
Oxford   New York   Toronto
Delhi   Bombay   Calcutta   Madras   Karachi
Kuala Lumpur   Singapore   Hong Kong   Tokyo
Nairobi   Dar es Salaam   Cape Town
Melbourne   Auckland   Madrid
and associated companies in
Berlin   Ibadan

Oxford is a trade mark of Oxford University Press

Published in the United States
by Oxford University Press Inc., New York

A catalogue record for this book is available from the British Library

Library of Congress Cataloging in Publication Data
(Data available)
ISBN 0 19 851775 0

Typeset by Cotswold Typesetting Ltd, Gloucester
Printed in Great Britain by
Biddles Ltd, Guildford and Kings Lynn

PREFACE

Scientific reductionism, often characterized as a method of understanding the whole by examining the parts, has served science and technology supremely well since the beginning of the nineteenth century. According to one of its most influential proponents, T. H. Huxley, it is also a perspective that takes nature and the universe to be deterministic, immutable, and non-anthropocentric; a perspective in which all biological and mental events are reducible to physical events and all physical events are reducible to properties of matter–energy. This austere outlook is widely interpreted by non-scientists to be a philosophy that reduces everything, including a belief in humanity; consequently reductionism has been instrumental in promoting the idea that there is a split between science and the rest of culture.

The wheel of science, however, continues to turn. Twentieth-century discoveries of new phenomena at successive levels in matter, living organisms, and mind–brain relationships have led to a more dynamic, emergent, relational view of nature. There is a new emphasis on holism, on an appreciation of nature's complex combinations of structure and openness, law and chance, order and chaos, determinism and probability. Nature, according to these new perspectives, is constituted by events and their relationships as much as by separate substances or separate particles. Historicity, moreover, is seen as an important characteristic of science; and science itself is conditioned by history. These trends are not new within the scientific community, but opinion about their validity and their significance varies markedly among scientists as between scientific disciplines.

Against this background a group of leading expositors of science, mathematics, and philosophy of science, met at Jesus College, Cambridge in September 1992 to discuss the continued primacy of reductionism as a key to understanding nature as we approach the twenty-first century. The participants included leading physicists, mathematicians, chemists, and biologists. A number of strongly conflicting 'scientific visions' (as Freeman Dyson would put it) were aired, including polarities of viewpoint as expressed in Peter Atkins's 'The limitless power of science' and Mary Midgley's energetic response, 'Reductive megalomania'.

Freeman Dyson, of the Institute for Advanced Study at Princeton, whose career as a physicist spans five decades, introduced the symposium by stressing the pluralistic nature of science, and by arguing that scientific practice is closer to an art or craft than an all-embracing philosophy. Science, he said, is 'an alliance of free spirits rebelling against the local culture', and its most significant advances arise from its new tools rather than new concepts. His insistence that scientific visions should remain humble in the face of what he calls 'nature's imagination' gave us the title for this published collection of the symposium's proceedings.

Roger Penrose and Gregory Chaitin, world-class mathematicians, argued that mathematics and physics could not be reductionist in the classic sense of the term. Penrose produced a series of impossible triangles, knots, and tangles, to demonstrate that their intrinsic 'impossibility' is part of the holistic structure of the shapes, making the notion of holism 'mathematically respectable'. Progressing to 'quantum entanglement' in the EPR paradox and Bell's theorem, Penrose argued that since particles are not independent one was forced to speculate on the 'connectedness of everything'. Scientific measurement, in fact, seemed to be a process of momentary 'disentangling'.

Reference to the work of Kurt Gödel and his great incompleteness theorem—described by Gregory Chaitin as the 'most profound result, the most mysterious result, in mathematics'—was a constant thread throughout the contributions. Both Penrose and Chaitin cited Gödel in arguing that mathematical belief cannot be limited to algorithmic mathematics, while Chaitin stressed the unpredictability of mathematics, demonstrating that a random sequence of bits of information is irreducible and 'algorithmically incompressible'. The astrophysicist John Barrow picked up this theme in his contribution 'Theories of Everything', arguing that science is a search for compressions, and concluding that the current quest for Theories of Everything might well be necessary, yet insufficient, to understand our Universe. We also require knowledge, he insisted, of initial conditions, forces, particles, constants, broken symmetry, selection effects, organizational principles, and even the categories of thought itself.

As the meeting developed there was an increasing preoccupation with the implications of human beings as observers in the universe, and the rapid advance of neuroscience in shedding light on consciousness and the mind–body problem. We agreed that the current quest to understand the mind reflects the most significant and altering perspective on reductionism, especially where psychology, philosophy of mind, and cognitive science meet the physical sciences. Gerald Edelman, Nobel laureate and neuroscientist, assisted by Giulio Tononi, eloquently demonstrated how neuroscience was leading towards the completion of the Darwinian Project and a shift away from the cognitive science view—'that minds are machines, and that nature is a piece of tape'. Oliver Sacks, the neurologist, followed this with a useful historical and cultural commentary on the significance of Edelman's theory for human individuality and autonomy.

Margaret Boden carried the human issue further in her discussion of the social implications of artificial intelligence, arguing that there was nothing to be feared from synthetic modelling of mental functions. Drawing a distinction between experience-dependent artefacts and programmed machines, she suggested that current artificial intelligence work left far more room for human dignity than behaviourism with its depersonalizing and mechanistic associations. Meanwhile, Hao Wang, in his paper on machines and souls, urged that we should not be unquestioning about psychophysical parallelism—the notion that minds and brains are equivalent—arguing with William James that parallelism is compatible with less strong forms of dependence of consciousness on the brain.

While the radical proponents of reductionism, both as a scientific method and as a philosophy, vigorously defended their corner, the majority of participants concurred with Gerald Edelman's spirited conclusion that the attempt to 'put mind back into nature' constitutes the end of the old Enlightenment and the beginning of the new. The 'neuroscience revolution', he maintained, was as historically significant as the Copernican revolution in cosmology—with crucial bearings on reductionism. Despite the success of reductionism in physics, chemistry, and molecular biology, 'it becomes silly reductionism when it is applied exclusively to the matter of the mind'. The workings of the mind, he went on, 'go beyond Newtonian causation. The workings of higher-order memories go beyond the description of temporal succession in physics. Finally, individual selfhood in society is to some extent a historical accident.'

The collected essays and papers that form this book are the result of the Jesus College symposium. Some were written specifically for the meeting, others are the result of material reworked subsequently in the light of our discussions. Patricia and Paul Churchland's contribution has been published elsewhere, and the substance of Gerald Edelman's postscript, as it appears here, is extracted from his book *Bright air, brilliant fire*, published after the meeting.

The symposium was made possible by the generosity of the Mrs L. D. Rope Third Charitable Settlement and the encouragement of Mr Crispin Rope. Thanks are also due to Jeremy Butterfield, Peter Lipton, Tomás Carruthers, and Bruce Wilcock, for valuable assistance. Finally I must thank the Master and Fellows of Jesus College, Cambridge, for their enthusiastic support.

Jesus College,   J.C.
Cambridge
July 1994

CONTENTS

CONTRIBUTORS

P. W. Atkins
Lecturer in Physical Chemistry in the University of Oxford. Author of *Molecular quantum mechanics* and *Creation revisited.*

John D. Barrow
Professor of Astronomy at the University of Sussex. Author of *The world within the world* and *Theories of Everything.*

Margaret A. Boden
Professor of Philosophy and Psychology in the School of Cognitive Sciences at the University of Sussex. Author of *The creative mind and artificial intelligence* and *Natural man.*

Gregory J. Chaitin
Member of IBM's Thomas J. Watson Research Center in New York.

Patricia S. Churchland
Professor of Philosophy in the University of California at San Diego. Author of *Neurophilosophy.*

Paul M. Churchland
Professor of Philosophy in the University of California at San Diego. Author of *Matter and consciousness.*

W. F. Clocksin
Lecturer in Computer Science at the University of Cambridge, and Fellow of Trinity Hall. Author (with C. S. Mellish) of *Programming in PROLOG.*

John Cornwell
Director of the Science and Human Dimension Project, Jesus College, Cambridge.

Freeman Dyson
Professor of Physics at the Institute for Advanced Study at Princeton. Author of *Disturbing the universe* and *Infinite in all directions.*

Gerald M. Edelman
Director of the Neurosciences Institute at San Diego, California, and Nobel Prize winner for Physiology and Medicine in 1972. Author of *Neural Darwinism.*

Mary Midgley
Formerly Lecturer in Philosophy in the University of Newcastle upon Tyne. Author of *Wisdom, information and wonder* and *Science as salvation*.

Roger Penrose
Rouse Ball Professor of Mathematics in the University of Oxford. Author of *The Emperor's new mind*.

Oliver Sacks
Professor of Neurology at the Albert Einstein Clinic in New York. Author of *The man who mistook his wife for a hat*.

Giulio Tononi
The Neurosciences Institute at San Diego, California.

Hao Wang
Professor of Logic at the Rockefeller University in New York. Author of *Reflections on Kurt Gödel*.

The scientist as rebel

FREEMAN DYSON

There is no such thing as a unique scientific vision, any more than there is a unique poetic vision. Science is a mosaic of partial and conflicting visions. But there is one common element in these visions. The common element is rebellion against the restrictions imposed by the locally prevailing culture, Western or Eastern as the case may be. The vision of science is not specifically Western. It is no more Western than it is Arab or Indian or Japanese or Chinese. Arabs and Indians and Japanese and Chinese had a big share in the development of modern science. And two thousand years earlier, the beginnings of ancient science were as much Babylonian and Egyptian as Greek. One of the central facts about science is that it pays no attention to East and West and North and South and black and yellow and white. It belongs to everybody who is willing to make the effort to learn it. And what is true of science is also true of poetry. Poetry was not invented by Westerners. India has poetry older than Homer. Poetry runs as deep in Arab and Japanese culture as it does in Russian and English. Just because I quote poems in English, it does not follow that the vision of poetry has to be Western. Poetry and science are gifts given to all of humanity.

For the great Arab mathematician and astronomer Omar Khayyám, science was a rebellion against the intellectual constraints of Islam, a rebellion which Khayyám expressed more directly in his incomparable verses:

> And that inverted bowl they call the sky,
> Whereunder crawling cooped we live and die,
>   Lift not your hands to *it* for help, for it
> As impotently rolls as you or I.

For the first generations of Japanese scientists in the nineteenth century, science was a rebellion against their traditional culture of feudalism. For the great Indian physicists of this century, Raman, Bose, and Saha, science was a double rebellion, first against English domination and second against the fatalistic ethic of Hinduism. And in the West too, great scientists from Galileo to Einstein have been rebels. Here is how Einstein himself described the situation:

When I was in the seventh grade at the Luitpold Gymnasium in Munich, I was summoned by my home-room teacher who expressed the wish that I leave the school. To my remark that I had done nothing amiss, he replied only, 'Your mere presence spoils the respect of the class for me'.

Einstein was glad to be helpful to the teacher. He followed the teacher's advice and dropped out of school at the age of fifteen.

From these and many other examples we see that science is not governed by the rules of Western philosophy or Western methodology. Science is an alliance of free spirits in all cultures rebelling against the local tyranny that each culture imposes on its children. In so far as I am a scientist, my vision of the universe is not reductionist or anti-reductionist. I have no use for Western isms of any kind. Like Loren Eiseley, I feel myself a traveller on a journey that is far longer than the history of nations and philosophies, longer even than the history of our species.

A few years ago an exhibition of Palaeolithic cave-art came to the Museum of Natural History in New York. It was a wonderful opportunity to see in one place the carvings in stone and bone that are normally kept in a dozen separate museums in France. Most of the carvings were done in France about fourteen thousand years ago, during a short flowering of artistic creation at the very end of the last Ice Age. The beauty and delicacy of the carving is extraordinary. The people who carved these objects cannot have been ordinary hunters amusing themselves in front of the cave-fire. They must have been trained artists sustained by a high culture. And the greatest surprise, when you see these objects for the first time, is the fact that their culture is not Western. They have no resemblance at all to the primitive art that arose ten thousand years later in Mesopotamia and Egypt and Crete. If I had not known that the old cave-art was found in France, I would have guessed that it came from Japan. The style looks today more Japanese than European. That exhibition showed us vividly that over periods of ten thousand years the distinctions between Western and Eastern and African culture lose all meaning. Over a time-span of a hundred thousand years we are all Africans. And over a time-span of three hundred million years we are all amphibians, waddling uncertainly out of dried-up ponds on to the alien and hostile land.

And with this long view of the past goes Robinson Jeffers's even longer view of the future. In the long view, not only European civilization but the human

species itself is transitory. Here is the vision of Robinson Jeffers, expressed in his
poem 'The double axe'.

> "Come, little ones.
> You are worth no more than the foxes and yellow wolfkins,
>   yet I will give you wisdom.
> O future children:
> Trouble is coming; the world as of the present time
> Sails on its rocks; but you will be born and live
> Afterwards. Also a day will come when the earth
> Will scratch herself and smile and rub off humanity;
> But you will be born before that.
>
> . . .
>
> "Time will come, no doubt,
> When the sun too shall die; the planets will freeze, and
>   the air on them; frozen gases, white flakes of air
> Will be the dust: which no wind ever will stir: this very
>   dust in dim starlight glistening
> Is dead wind, the white corpse of wind.
> Also the galaxy will die; the glitter of the Milky Way,
>   our universe, all the stars that have names are dead.
> Vast is the night. How you have grown, dear night,
>   walking your empty halls, how tall!"

Robinson Jeffers was no scientist, but he expressed better than any other poet
the scientist's vision. Ironic, detached, contemptuous like Einstein of national
pride and cultural taboos, he stood in awe of nature alone. He stood alone in
uncompromising opposition to the follies of the Second World War. His poems
during those years of patriotic frenzy were unpublishable. 'The double axe'
was finally published in 1948, after a long dispute between Jeffers and his
editors. I discovered Jeffers thirty years later, when the sadness and the passion
of the war had become a distant memory. Fortunately, his works are now in
print and you can read them for yourselves.

Science as subversion has a long history. There is a long list of scientists who
sat in jail and other scientists who helped get them out and incidentally saved
their lives. In our own century we have seen the physicist Landau sitting in jail
in the Soviet Union and Kapitsa risking his own life by appealing to Stalin to let
Landau out. We have seen the mathematician André Weil sitting in jail in
Finland during the Winter War of 1939–40 and Lars Ahlfors saving his life.
The finest moment in the history of the Institute for Advanced Study, where I
work, came in 1957, when we appointed the mathematician Chandler Davis a

member of the Institute, with financial support provided by the American government through the National Science Foundation. Chandler was then a convicted felon because he refused to rat on his friends when questioned by the House Unamerican Activities Committee. He had been convicted of Contempt of Congress for not answering questions and had appealed against his conviction to the Supreme Court. While his case was under appeal, he came to Princeton and continued doing mathematics. That is a good example of science as subversion. After his Institute fellowship was over, he lost his appeal and sat for six months in jail. Chandler is now a distinguished professor at the University of Toronto and is actively engaged in helping people in jail to get out. Another example of science as subversion is Andrei Sakharov. Chandler Davis and Sakharov belong to an old tradition in science that goes all the way back to the rebels Franklin and Priestley in the eighteenth century, to Galileo and Giordano Bruno in the seventeenth and sixteenth. If science ceases to be a rebellion against authority, then it does not deserve the talents of our brightest children. I was lucky to be introduced to science at school as a subversive activity of the younger boys. We organized a Science Society as an act of rebellion against compulsory Latin and compulsory football. We should try to introduce our children to science today as a rebellion against poverty and ugliness and militarism and economic injustice.

The vision of science as rebellion was articulated here in Cambridge with great clarity on 4 February 1923, in a lecture by the biologist J. B. S. Haldane to the Society of Heretics. The lecture was published as a little book with the title *Daedalus*. Here is Haldane's vision of the role of the scientist. I have taken the liberty to abbreviate Haldane slightly and to omit the phrases that he quoted in Latin and Greek, since unfortunately I can no longer assume that the heretics of Cambridge are fluent in those languages.

The conservative has but little to fear from the man whose reason is the servant of his passions, but let him beware of him in whom reason has become the greatest and most terrible of the passions. These are the wreckers of outworn empires and civilizations, doubters, disintegrators, deicides. In the past they have been men . . . like Voltaire, Bentham, Thales, Marx, . . . but I think that Darwin furnishes an example of the same relentlessness of reason in the field of science. I suspect that as it becomes clear that at present reason not only has a freer play in science than elsewhere, but can produce as great effects on the world through science as through politics, philosophy or literature, there will be more Darwins . . .

We must regard science, then, from three points of view. First, it is the free activity of man's divine faculties of reason and imagination. Secondly, it is the answer of the few to the demands of the many for wealth, comfort and victory, . . . gifts which it will grant only in exchange for peace, security and stagnation. Finally it is man's gradual conquest, first of space and time, then of matter as such, then of his own body and those of other living beings, and finally the subjugation of the dark and evil elements in his own soul.

I have already made it clear that I have a low opinion of reductionism, which seems to me to be at best irrelevant and at worst misleading as a description of what science is about. Let me begin with pure mathematics. Here the failure of reductionism has been demonstrated by rigorous proof. This will be a familiar story to many of you. The great mathematician David Hilbert, after thirty years of high creative achievement on the frontiers of mathematics, walked into the blind alley of reductionism. In his later years he espoused a programme of formalization, which aimed to reduce the whole of mathematics to a collection of formal statements using a finite alphabet of symbols and a finite set of axioms and rules of inference. This was reductionism in the most literal sense, reducing mathematics to a set of marks written on paper, and deliberately ignoring the context of ideas and applications that give meaning to the marks. Hilbert then proposed to solve the problems of mathematics by finding a general decision-process that could decide, given any formal statement composed of mathematical symbols, whether that statement was true or false. He called the problem of finding this decision-process the *Entscheidungsproblem*. He dreamed of solving the *Entscheidungsproblem* and thereby solving as corollaries all the famous unsolved problems of mathematics. This was to be the crowning achievement of his life, the achievement that would outshine all the achievements of earlier mathematicians who solved problems only one at a time.

The essence of Hilbert's programme was to find a decision process that would operate on symbols in a purely mechanical fashion, without requiring any understanding of their meaning. Since mathematics was reduced to a collection of marks on paper, the decision process should concern itself only with the marks and not with the fallible human intuitions out of which the marks were reduced. In spite of prolonged efforts of Hilbert and his disciples, the *Entscheidungsproblem* was never solved. Success was achieved only in highly restricted domains of mathematics, excluding all the deeper and more interesting concepts. Hilbert never gave up hope, but as the years went by his programme became an exercise in formal logic having little connection with real mathematics. Finally, when Hilbert was seventy years old, Kurt Gödel proved by a brilliant analysis that the *Entscheidungsproblem* as Hilbert formulated it cannot be solved. Gödel proved that in any formalization of mathematics including the rules of ordinary arithmetic a formal process for separating statements into true and false cannot exist. He proved the stronger result which is now known as Gödel's Theorem, that in any formalization of mathematics including the rules of ordinary arithmetic there are meaningful arithmetical statements that cannot be proved true or false. Gödel's Theorem shows conclusively that in pure mathematics reductionism does not work. To decide whether a mathematical statement is true, it is not sufficient to reduce the statement to marks on paper and to study the behaviour of the marks. Except in trivial cases, you can decide the truth of a statement only by studying its meaning and its context in the larger world of mathematical ideas.

It is a curious paradox that several of the greatest and most creative spirits in science, after achieving important discoveries by following their unfettered imaginations, were in their later years obsessed with reductionist philosophy and as a result became sterile. Hilbert was a prime example of this paradox. Einstein was another. Like Hilbert, Einstein did his great work up to the age of forty without any reductionist bias. His crowning achievement, the general relativistic theory of gravitation, grew out of a deep physical understanding of natural processes. Only at the very end of his ten-year struggle to understand gravitation did he reduce the outcome of his understanding to a finite set of field-equations. But like Hilbert, as he grew older he concentrated his attention more and more on the formal properties of his equations, and he lost interest in the wider universe of ideas out of which the equations arose. His last twenty years were spent in a fruitless search for a set of equations that would unify the whole of physics, without paying attention to the rapidly proliferating experimental discoveries that any unified theory would finally have to explain. I do not need to say more about this tragic and well-known story of Einstein's lonely attempt to reduce physics to a finite set of marks on paper. His attempt failed as dismally as Hilbert's attempt to do the same thing with mathematics. I shall instead discuss another aspect of Einstein's later life, an aspect that has received less attention than his quest for the unifield field equations: his extraordinary hostility to the idea of black holes.

Black holes were invented by Oppenheimer and Snyder in 1940. Starting from Einstein's theory of general relativity, Oppenheimer and Snyder found solutions of Einstein's equations describing what happens to a massive star when it has exhausted its supplies of nuclear energy. The star collapses gravitationally and disappears from the visible universe, leaving behind only an intense gravitational field to mark its presence. The star remains in a state of permanent free fall, collapsing endlessly inward into the gravitational pit without ever reaching the bottom. This solution of Einstein's equations was profoundly novel. It has had enormous impact on the later development of astrophysics. We now know that black holes ranging in mass from a few suns to a few billion suns actually exist and play a dominant role in the economy of the universe. In my opinion, the black hole is incomparably the most exciting and the most important consequence of general relativity. Black holes are the places in the universe where general relativity is decisive. But Einstein never acknowledged his brainchild. Einstein was not merely sceptical, he was actively hostile to the idea of black holes. He thought that the black-hole solution was a blemish to be removed from his theory by a better mathematical formulation, not a consequence to be tested by observation. He never expressed the slightest enthusiasm for black holes, either as a concept or as a physical possibility. Oddly enough, Oppenheimer too in later life was uninterested in black holes, although in retrospect we can say that they were his most important contribution to science. The older Einstein and the older Oppenheimer were blind to the mathematical beauty of black holes, and indifferent to the question whether black holes actually exist.

How did this blindness and this indifference come about? I never discussed this question directly with Einstein, but I discussed it several times with Oppenheimer and I believe that Oppenheimer's answer applies equally to Einstein. Oppenheimer in his later years believed that the only problem worthy of the attention of a serious theoretical physicist was the discovery of the fundamental equations of physics. Einstein certainly felt the same way. To discover the right equations was all that mattered. Once you had discovered the right equations, then the study of particular solutions of the equations would be a routine exercise for second-rate physicists or graduate students. In Oppenheimer's view, it would be a waste of his precious time, or of mine, to concern ourselves with the details of particular solutions. This was how the philosophy of reductionism led Oppenheimer and Einstein astray. Since the only purpose of physics was to reduce the world of physical phenomena to a finite set of fundamental equations, the study of particular solutions such as black holes was an undesirable distraction from the central goal. Like Hilbert, they were not content to solve particular problems one at a time. They were entranced by the dream of solving all the basic problems at once. And as a result, they failed in their later years to solve any problems at all.

In the history of science it happens not infrequently that a reductionist approach leads to spectacular success. Frequently the understanding of a complicated system as a whole is impossible without an understanding of its component parts. And sometimes the understanding of a whole field of science is suddenly advanced by the discovery of a single basic equation. Thus it happened that the Schrödinger equation in 1926 and the Dirac equation in 1927 brought a miraculous order into the previously mysterious processes of atomic physics. The equations of Schrödinger and Dirac were triumphs of reductionism. Bewildering complexities of chemistry and physics were reduced to two lines of algebraic symbols. These triumphs were in Oppenheimer's mind when he belittled his own discovery of black holes. Compared with the abstract beauty and simplicity of the Dirac equation, the black-hole solution seemed to him ugly, complicated, and lacking in fundamental significance.

But it happens at least equally often in the history of science that the understanding of the component parts of a composite system is impossible without an understanding of the behaviour of the system as a whole. And it often happens that the understanding of the mathematical nature of an equation is impossible without a detailed understanding of its solutions. The black hole is a case in point. One could say without exaggeration that Einstein's equations of general relativity were understood only at a very superficial level before the discovery of the black hole. During the fifty years since the black hole was invented, a deep mathematical understanding of the geometrical structure of space–time has slowly emerged, with the black-hole solution playing a fundamental role in the structure. The progress of science requires the growth of understanding in both directions, downward from the whole to the parts and upward from the parts to the whole. A reductionist philosophy, arbitrarily proclaiming that the growth of understanding must go only in one direction, makes no scientific sense. Indeed, dogmatic philosophical beliefs of any kind have no place in science.

Science in its everyday practice is much closer to art than to philosophy. When I look at Gödel's proof of his undecidability theorem, I do not see a philosophical argument. The proof is a soaring piece of architecture, as unique and as lovely as Chartres cathedral. Gödel took Hilbert's formalized axioms of mathematics as his building-blocks and built out of them a lofty structure of ideas into which he could finally insert his undecidable arithmetical statement as the keystone of the arch. The proof is a great work of art. It is a construction, not a reduction. It destroyed Hilbert's dream of reducing all mathematics to a few equations, and replaced it with a greater dream of mathematics as an endlessly growing realm of ideas. Gödel proved that in mathematics the whole is always greater than the sum of the parts. Every formalization of mathematics raises questions that reach beyond the limits of the formalism into unexplored territory.

The black-hole solution of Einstein's equations is also a work of art. The black hole is not as majestic as Gödel's proof, but it has the essential features of a work of art: uniqueness, beauty, and unexpectedness. Oppenheimer and Snyder built out of Einstein's equations a structure that Einstein had never imagined. The idea of matter in permanent free fall was hidden in the equations, but nobody saw it until it was revealed in the Oppenheimer–Snyder solution. On a much more humble level, my own activities as a theoretical physicist have a similar quality. When I am working, I feel myself to be practising a craft rather than following a method. When I did my most important piece of work as a young man, putting together the ideas of Tomonaga, Schwinger, and Feynman to obtain a simplified version of quantum electrodynamics, I had consciously in mind a metaphor to describe what I was doing. The metaphor was bridge-building. Tomonaga and Schwinger had built solid foundations on one side of a river of ignorance, Feynman had built solid foundations on the other side, and my job was to design and build the cantilevers reaching out over the water until they met in the middle. The metaphor was a good one. The bridge that I built is still serviceable and still carrying traffic forty years later. The same metaphor describes well the greater work of unification achieved by Weinberg and Salam when they bridged the gap between electrodynamics and the weak interactions. In each case, after the work of unification is done, the whole stands higher than the parts.

In recent years there has been great dispute among historians of science, some believing that science is driven by social forces, others believing that science transcends social forces and is driven by its own internal logic and by the objective facts of nature. Historians of the first group write social history, those of the second group write intellectual history. Since I believe that scientists should be artists and rebels, obeying their own instincts rather than social demands or philosophical principles, I do not fully agree with either view of history. Nevertheless, scientists should pay attention to the historians. We have much to learn, especially from the social historians.

Many years ago, when I was in Zürich, I went to see the play *The physicists* by the Swiss playwright Dürrenmatt. The characters in the play are grotesque caricatures, wearing the costumes and using the names of Newton, Einstein, and Möbius. The action takes place in a lunatic asylum where the physicists are patients. In the first act they entertain themselves by murdering their nurses, and in the second act they are revealed to be secret agents in the pay of rival intelligence services. I found the play amusing but at the same time irritating. These absurd creatures on the stage had no resemblance at all to any real physicist. I complained about the unreality of the characters to my friend Markus Fierz, a well-known Swiss physicist who came with me to the play. 'But don't you see?', said Fierz, 'The whole point of the play is to show us how we look to the rest of the human race.' Fierz was right. The image of noble and virtuous dedication to truth, the image that scientists have traditionally presented to the public, is no longer credible. The public, having found out that the traditional image of the scientist as a secular saint is false, has gone to the opposite extreme and imagines us to be irresponsible devils playing with human lives. Dürrenmatt has held up the mirror to us and has shown us the image of ourselves as the public sees us. It is our task now to dispel these fantasies with facts, showing to the public that scientists are neither saints nor devils but human beings sharing the common weaknesses of our species.

Historians who believe in the transcendence of science have portrayed scientists as living in a transcendent world of the intellect, superior to the transient, corruptible, mundane realities of the social world. Any scientist who claims to follow such exalted ideals is easily held up to ridicule as a pious fraud. We all know that scientists, like television evangelists and politicians, are not immune to the corrupting influences of power and money. Much of the history of science, like the history of religion, is a history of struggles driven by power and money. And yet, this is not the whole story. Genuine saints occasionally play an important role, both in religion and in science. Einstein was an important figure in the history of science, and he was a firm believer in transcendence. For Einstein, science as a way of escape from mundane reality was no pretence. For many scientists less divinely gifted than Einstein, the chief reward for being a scientist is not the power and the money but the chance of catching a glimpse of the transcendent beauty of nature.

Both in science and in history there is room for a variety of styles and purposes. There is no necessary contradiction between the transcendence of science and the realities of social history. One may believe that in science nature will ultimately have the last word, and still recognize an enormous role for human vainglory and viciousness in the practice of science before the last word is spoken. One may believe that the historian's job is to expose the hidden influences of power and money, and still recognize that the laws of nature cannot be bent and cannot be corrupted by power and money. To my mind, the history of science is most illuminating when the frailties of human actors are put into juxtaposition with the transcendence of nature's laws.

Francis Crick is one of the great scientists of our century. He has recently published his personal narrative of the microbiological revolution that he helped to bring about, with a title borrowed from Keats, *What mad pursuit*. One of the most illuminating passages in his account compares two discoveries in which he was involved. One was the discovery of the double-helix structure of DNA, the other was the discovery of the triple-helix structure of the collagen molecule. Both molecules are biologically important, DNA being the carrier of genetic information, collagen being the protein that holds human bodies together. The two discoveries involved similar scientific techniques and aroused similar competitive passions in the scientists racing to be the first to find the structure. Crick says that the two discoveries caused him equal excitement and equal pleasure at the time he was working on them. From the point of view of a historian who believes that science is a purely social construction, the two discoveries should have been equally significant. But in history as Crick experienced it, the two helices were not equal. The double helix became the driving-force of a new science, while the triple helix remained a footnote of interest only to specialists. Crick asks the question, How are the different fates of the two helices to be explained? He answers the question by saying that human and social influences cannot explain the difference, that only the transcendent beauty of the double helix structure and its genetic function can explain the difference. Nature herself, and not the scientist, decided what was important. In the history of the double helix, transcendence was real. Crick gives himself the credit for choosing an important problem to work on, but, he says, only nature herself could tell how transcendentally important it would turn out to be.

My message is that science is a human activity, and the best way to understand it is to understand the individual human beings who practise it. Science is an art form and not a philosophical method. The great advances in science usually result from new tools rather than from new doctrines. If we try to squeeze science into a single philosophical viewpoint such as reductionism, we are like Procrustes chopping off the feet of his guests when they do not fit on to his bed. Science flourishes best when it uses freely all the tools at hand, unconstrained by preconceived notions of what science ought to be. Every time we introduce a new tool, it always leads to new and unexpected discoveries, because Nature's imagination is richer than ours.

CHAPTER TWO

Must mathematical physics be reductionist?

ROGER PENROSE

I think that one of the things to be learned from Freeman Dyson's presentation is that one cannot be a self-respecting scientist without also being a rebel. This would seem to imply that in this book we are urged to rebel against the point of view of reductionism. This is difficult to do unless we know first what reductionism is; and in fact I am not clear that I do know what the word means. Some forms of reductionism are almost synonymous with 'scientific', in which case we find ourselves driven to a contradiction. If we are going to be scientific, we have to be rebellious, but that rebelliousness leads us to resist the critical stance of the symposium—which would mean science itself if we take 'reductionist' to mean merely 'scientific'.

I am going to interpret the word 'reductionism' in various ways that are different from this. I have certainly been accused of being non-reductionist, and I have to try to understand what this charge means. There are perhaps two main alternative interpretations that come to mind, and I shall try to illustrate them in this contribution.

There are indeed certain areas of mathematics that are, in one real sense, non-reductionist. This does not prevent us from discussing problems in those areas in a very clear and characteristically mathematical way. We might take reductionism to mean the breaking-down of things into smaller and smaller parts, so that if you understand how the small parts work, that will in principle tell you how the big thing works. It may be that you need other kinds of ideas to say interesting things about the big things, as Freeman Dyson stressed, but here, the idea of reductionism is that the behaviour of the big things we study is governed by the behaviour of the individual units of which they are composed.

That is one strand of reductionism. Another strand of reductionism may be related to this by asking if knowledge of the present behaviour of those small units will allow us to predict the behaviour of the big things in the future? Determinism has been a central concern of philosophy and physics for many centuries: we may understand it as the proposition that the behaviour of the system at one time will determine what it does at a later time. It is important to emphasize that the *computability* of the behaviour at later times is a separate question. I want to stress the difference between computability and determinism because the distinction may prove to be an important one. Before we address it, I should like to return to our first interpretation of reductionism and illustrate how it is possible to talk about holistic concepts in a clear and mathematically precise fashion.

HOLISM AS MATHEMATICS

Let us start by investigating what we mean by a holistic concept. Figure 2.1 illustrates an impossible triangle. By virtue of what particular property is the triangle impossible? The fact that it possesses this feature of impossibility is clear. It is supposed to be an image that conveys to the mind a particular three-dimensional object, but that three-dimensional object simply cannot exist in ordinary space. We must ask ourselves what is wrong with the picture: can we point to somewhere in the picture where the mistake was made?

We might locate the impossibility in one specified corner of the triangle, so that if we covered up that corner the figure would make sense as the

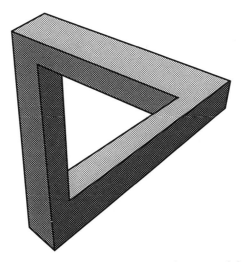

Fig. 2.1. An impossible triangle. Where is the impossibility?

representation of a possible three-dimensional object. We could then say that the impossibility has disappeared. Apparently it did not reside in the rest of the picture. If we hide *any* one corner, or remove *any* edge, the structure suddenly becomes possible. When we study the complete picture, however, the impossibility re-emerges.

This impossibility is a property of the whole structure. It cannot be localized in some part: it is a feature of the complete mathematical object, so it is a *holistic* property. There are areas of mathematics dedicated to discussing this kind of property in a rigorous way.

Let us imagine breaking the figure into three pieces, as illustrated (Fig. 2.2), and then glueing the parts together again. We could continue breaking the structure down, and each of the pieces would be a possible object in ordinary space. The process of glueing them together is the operation which eventually produces the impossible triangle. In technical mathematical terms, we take all the pieces together with all the specific glueing operations and by a factoring-out procedure, we extract what is called the cohomology. This mathematical concept abstracts what it is that interests us: the measure of impossibility of the triangle. In this particular instance, we can give this measure as a single real number, describing the 'degree of impossibility' of the triangle.

If I break this structure somewhere, so that it becomes a possible three-dimensional object, I can measure the ratio of the distances from my eye of one end of the break to the other end of the break. That ratio is a measure of the

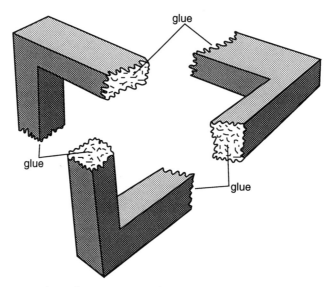

Fig. 2.2. Cohomology: an impossible triangle broken and reassembled.

impossibility of the structure and it will be the same, no matter where I cut the figure. This holds also with more than one break, no matter how many pieces I reduce it into, provided that I multiply all the relevant factors together. That perhaps gives you something of the idea of cohomology. In this particular structure, cohomology supplies the degree of impossibility of the object, and it is a perfectly well-defined and precise holistic mathematical concept.

If we take a knot (Fig. 2.3), we might ask another question: where does the property of knottedness reside in this knot? Like the impossibility of the triangle, we find that the 'knottedness' cannot be localized. It is a property of the structure as a whole. There are profound mathematical theories that deal with this question of knottedness: and though it looks simple enough, knottedness is more difficult to deal with than cohomology. Research in this field dates from the turn of the century and more recently and is now very well developed.

Another example is a Möbius band (Fig. 2.4). Where, we might ask, is the twist? If we take a piece of paper, twist it once and glue it together, we get a Möbius band. Then wherever we cut it, the twist is gone. It does not matter

Fig. 2.3. Where does the knottedness reside in the knot?

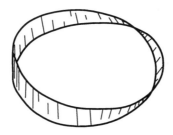

Fig. 2.4. A Möbius band. Where is the twist?

where we cut it, because once again the twist is a feature of the global structure and not something that can be localized. Once again, mathematics offers quite involved theories arising from the study of these twisted structures (called fibre bundles).

Finally we have (Fig. 2.5) a surface with a certain number of handles attached to it. Two such surfaces are said to be topologically equivalent if they can be deformed continuously one into the other. Where, we might ask, is the topology? We find that we cannot point to anywhere on the surface where we can localize it. The topology is a property of the global structure.

All these figures represent concepts important to modern physics. The idea that particles might be knots in some way goes back to Lord Kelvin. Such ideas have been revived and modernized and remain prominent in current thinking. Fibre bundles are crucial to the modern gauge theories of particle interactions. Surfaces like the one we examined are central to string theory.

Can we describe this sort of thing mathematically? There are in fact well-defined and often very subtle mathematical procedures for discussing holistic properties of the kind that I have illustrated above. There is nothing mystical about them. Holism is an entirely respectable, perfectly clear concept within modern mathematics.

QUANTUM NONLOCALITY

The objects we looked at above were mathematical, but I pointed out that they played an important role in some theories of particle physics. Moreover, holism

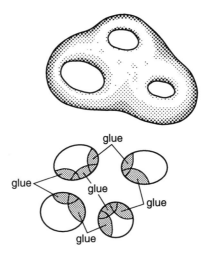

Fig. 2.5. A surface with handles. Where is the topology?

has a recognized place even in classical physics, which accepts that certain concepts are not localizable. The concept of energy in Einstein's general relativity theory, in particular, turns out to be non-local. It is the localization of the *gravitational* contribution to the energy in a system that turns out to be impossible. In general relativity, the energy and therefore (by Einstein's $E = mc^2$) also the *mass* of a system is not something we can pin down at any particular location. It has already become a holistic concept. With quantum mechanics, physics becomes holistic in an even more mysterious way.

Let us consider the phenomenon known as quantum entanglement. It was Schrödinger who first identified it, but Einstein, Podolsky, and Rosen pointed out the absurd nature of some of its implications—called EPR phenomena— and it was John Bell who showed, in effect, that quantum entanglement could be tested experimentally. The most famous such experiment was performed by Alain Aspect and his colleagues in Paris. We can describe it roughly in the following way. We have a number of atoms in one energy level, which from time to time drop down to a lower energy level. In the process, each emits a pair of photons. These photons go off in opposite directions and can be received by detectors, some 12 metres apart, that measure the photons' polarizations. The decision as to which of two possible directions is chosen in which to measure each photon's polarization is not made until they are in flight. Bell showed that any scheme in which these two photons are treated as separate independent objects would necessarily involve probabilities that satisfied certain laws. Quantum mechanics, by contrast, tells us that the probabilities will *not* satisfy such laws. The results of the experiment violated those laws in the way that quantum mechanics had predicted: so the quantum laws were vindicated. A quantum system like this, with two photons travelling a distance of 12 metres, reveals that the pair of photons cannot be regarded as consisting of two independent, separate things. This underwrites the essentially holistic nature of quantum systems. We usually cannot break down such systems into smaller and smaller units. The assumption that the photons in this experiment are independent units is not borne out by the results.

These experiments depend upon the measurement of joint probabilities. However, there has been interesting recent work of a similar nature, but which does *not* involve any probabilities (other than 0 or 1). This new work is associated with the names of Greenberger, Zeilinger, Horne, Shimony, Redhead, Clifton, and Butterfield. It concerns experiments involving three separated units, or perhaps four, instead of two. Now there are no probabilities at all: the puzzling EPR effects are obtained in simple yes/no constraints. If the objects were separate and independent entities, one would have to say that certain things could not happen, but quantum mechanics says that they *do* happen. The clear expectations are that it is quantum mechanics which is true, at least at this level, and the 'common-sense' view that widely separated objects must behave independently which is false.

There is a particular instance of Bell nonlocality without probability that I should like to discuss: I call it the *magic dodecahedra*. What we have to imagine is that here on Earth I receive a parcel sent from the far distant place where these things are produced, and an identical parcel is sent to a colleague of mine on Alpha Centauri. The package containing each magic dodecahedron comes with instructions telling us how these things behave. I should point out at this stage that the manufacturers of the dodecahedra—on the distant star Betelgeuse—have been producing these things for millions of years and they have never made a mistake. We can rely absolutely on the claims of the manufacturers when we perform our experiment.

Each of us selects a vertex of the dodecahedra and then takes three vertices adjacent to the one selected, as illustrated in Fig. 2.6. There are buttons on each of the vertices, and we press these three buttons in random order. A bell may ring; but it is going to ring on only one of these vertices and it may not ring on any of them. In the instructions I received with the dodecahedron I was told that certain things never occur, and these instructions relate what happens with my dodecahedron to what happens with the one received by my colleague on Alpha Centauri. First, if the bell rings on one of the vertices of my dodecahedron, then the bell cannot fail to ring on the opposite vertex on the other dodecahedron if my colleague presses that one, and vice versa. Thus, if my colleague happens to press a vertex on his dodecahedron directly opposite the button I press, whether it is the first, the second, or the third of the three adjacent to my selected one, then if the bell has rung for me, it must also ring for him and vice versa. Secondly, if he happens to select the same vertex as I do, and presses the three buttons adjacent to it, the instructions tell me that it

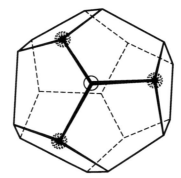

Fig. 2.6. A magic dodecahedron. The ringed vertex is selected and the three dotted vertices are pressed.

cannot be the case that the bell fails to ring for those three buttons here and also for the corresponding three on Alpha Centauri. Of course, I shall not know what has happened with his dodecahedron for a while, because my colleague is four light years away from Earth. However, since the makers of these dodecahedra have not been wrong about their behaviour in a million years, we must presumably just accept their rules.

The only problem is that there turns out to be something rather contradictory in what they say. The first rule states that if I press a button on the corner of my dodecahedron here on Earth, and the bell rings, it must then ring for the opposite button on the dodecahedron on Alpha Centauri. The makers on Betelgeuse are not to know whether my colleague on Alpha Centauri is going to choose that opposite vertex as one of the combination of buttons he presses. If my bell rings, his rings, and if my bell does not ring, his does not. The ringing or otherwise of my bell seems to be something that my dodecahedron must be prepared to do ahead of time, since the manufacturers cannot guarantee that my colleague on Alpha Centauri will not press the button opposite to mine. Our instructions require that both dodecahedra are to be carefully orientated so that the vertices can be related to each other, and it seems that my dodecahedron knows in advance whether or not to ring the bell if I press a particular button. I deduce that if I press two next-to-adjacent buttons, the bell cannot ring on both of them. Secondly, if it does not ring for the three buttons adjacent to my selected one—i.e. on three mutually next-to-adjacent ones—then it must ring on one of the opposite three buttons (see Fig. 2.5).

Taking these two properties which we seem to have conclusively derived, and doing a little work, we actually find that it is *impossible* to designate all the vertices of this dodecahedron as bell-ringers or silent, in accordance with the original guarantees of the makers. It just cannot be done. We might react by suggesting that the makers on Betelgeuse have finally made a mistake after a million years, but if we wait four light years for the message from my colleague on Alpha Centauri, we shall find that his results *do* agree with the guarantees we originally received. The magic dodecahedra are actually a curious example of *quantum entanglement*. In the above reasoning, I made the seemingly unexceptionable assumption that the two dodecahedra were *independent* objects. This assumption was false. Quantum entanglement—a genuinely holistic feature of quantum systems—entails that the objects in question are actually mysteriously connected, no matter how far apart they might be.

I shall not explain in detail how these dodecahedra are built: that is the province of those who know the appropriate physics. What I can say is that on Betelgeuse they had a spin-0 system which was split into two atoms, each of spin-$\frac{3}{2}$. At the centre of each dodecahedron is placed one of these atoms of

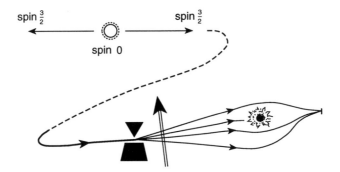

spin $\frac{3}{2}$ spin $\frac{3}{2}$

spin 0

Fig. 2.7. A spin-0 system that is split into two atoms, each of spin-$\frac{3}{2}$.

spin-$\frac{3}{2}$, and if I press a button on one of the vertices, that effects a measurement in the direction of that vertex. There are four possible outcomes. The atom chooses one of those four, and if it is in the second from top (to the experts: $m = \frac{1}{2}$) the bell rings (see Fig. 2.7). If not, the bell does not ring, in which case the remaining three possibilities are reassembled and we try again in another direction. With present technology, it might even be possible to set this up and perform an experiment of this nature.

According to standard theory and observed experimental fact, quantum entanglement certainly does occur. Widely separated physical objects are not independent of one another. I should also emphasize that according to standard quantum-mechanical theory, one cannot use this absence of independence to send a message faster than the speed of light from one object to the other. Nevertheless, the way in which the dodecahedra behave cannot be understood if we insist on regarding them as separate and independent objects. This is indeed an instance of the phenomenon of quantum entanglement. The phenomenon generally is prominent in many quantum systems. This particular example neatly illustrates that physical behaviour cannot be explained if we insist on treating the units involved as separate, independent local objects.

Why isn't everything entangled? If these two particles, four light years away from each other are entangled, why isn't the whole universe an extraordinary entangled mess? How can we deduce anything which has the semblance of locality? After all, if this 'local' form of reductionism is not right, and the universe cannot be separated into these independent pieces, why does our physics work so well?

The answer is not altogether clear, but it is noteworthy that in our example of the magic dodecahedra there was a process of *measurement*. Pressing the button carried out a measurement, and measurement seems to disentangle.

Once you perform a measurement you seem to have released this unit from its partner in some mysterious way. Perhaps we should ask what a measurement is, in physical terms.

Quantum theory is all about measurements but it never clarifies what a measurement actually is. In some sense, a measurement involves magnifying a quantum effect to something on a microscopic scale, so that we can see what is going on in the system. At the quantum level you might have to superpose all the possible things that might happen; once you have magnified to a macro level, only one thing or another actually happens, and not all these strange superpositions of alternatives happen all at once, which is what we must consider occurs at the quantum level.

Some would say that measurement occurs when the quantum system becomes involved with the environment in some sufficiently random way. Although the systems do not disentangle in the strict sense, they are *in effect* disentangled—or so it is claimed. We are provisionally able to treat the system as though it were not entangled, even though, in a strict sense according to standard theory, they would always remain entangled.

My own view is that measurements really *do* disentangle quantum systems: strictly speaking, this would require a change in fundamental physical theory. It is deeply unsatisfactory to have different rules applying in physics at the quantum level and at the classical level. Atoms and molecules seem to obey one set of rules which do not obtain at the classical level. Moving from one level to the other is difficult to do intelligibly: but here we are touching upon a subject of endless controversy. I had better not press the matter further here, but just make note of my own belief in the need for a new underlying theory, and move on.

HOLISM IN BRAIN ACTION?

Are quantum effects such as nonlocality and entanglement relevant to explanations of brain action? When I wrote *The Emperor's new mind*, I found it difficult to believe that nerve transmission could somehow preserve quantum entanglements and quantum superpositions, but I could not think of anything else. Nerve activity seems to be a large-scale, classical affair where each signal disturbs the environment so much that it is hard to see how quantum entanglements and superpositions could be maintained. Since then I have learned of another much more promising possibility from Stuart Hameroff, and I should like to outline it.

There is a good deal of computational activity below the level of neurones,

Hameroff argues, and he points to single-cell organisms such as amoebae and paramecia, which are capable of behaving in elaborate ways despite having no nervous system at all. They do, however, possess a structure known as a cytoskeleton, which not only governs their shape but also appears to control how they behave. A paramecium can swim towards its food, or away from danger. It can negotiate obstacles and even learn by experience. Clearly it is capable of complex and sophisticated activity; yet it does not possess a single neurone. If it is indeed the cytoskeleton that controls the behaviour of a paramecium, then we may expect cytoskeletons to be important in the behaviour of other cells. In fact, all eukaryotic cells possess cytoskeletons—including individual neurones.

If we study the structure of a cytoskeleton, we find tube-like threads known as microtubules, made out of arrays of protein molecules called tubulin dimers. These structures, it seems to me, are very plausibly regarded as quantum-mechanical rather than classical-level objects. There could be effects travelling down these tubes that do not disturb the environment, and these should then be understood in quantum rather than classical terms, where quantum entanglements and superpositions are preserved. It would be difficult to tell what is actually going on within such a system, but it would be quite different from the action of a nerve. Hameroff and his colleagues believe that messages may be travelling up and down these microtubules which represent basic computational activity, and that these messages govern the behaviour of one-celled organisms. Moreover, the activity in the microtubules within neurones can apparently influence or control the synaptic connections between neurones. Though microtubules are not concerned with the actual nerve signals, they may well govern the strength of the synapses.

It seems to me to be an interesting hypothesis that collective quantum effects may be taking place in these structures. At least part of the role of the microtubules in neurones is that they are significant in *maintaining* the strength of the synapse. Without the maintenance, the strength of the connection would decay. If quantum non-local effects are important in the explanation of brain activity, this seems to be the place to start looking for them. It is worth recalling that some people have objected to any kind of quantum explanation of brain activity on the grounds that the brain is too hot for quantum effects to be important. This may be compared with the collective quantum phenomenon of superconductivity, which was originally thought to occur only at temperatures of a few degrees above absolute zero. Superconductivity is now observed at comparatively high temperatures, and I would argue by analogy that the high temperature of the brain does not rule out the occurrence of collective quantum effects that might be relevant to brain activity.

Classical physics seems incapable of explaining a phenomenon so deeply mysterious as consciousness. The physical world we know is extremely mysterious, and the more we probe its foundations the more mysterious it becomes. That mystery does not mean that the world is irrational—that it is not susceptible to scientific understanding—but it does compel us to accept mystery, and these non-local quantum phenomena are certainly mysterious.

Brain activity is often explained by drawing parallels with the operation of computers, so let us consider the possible operation of quantum computers. Hameroff and his colleagues argue that computational activity is taking place in microtubules far below the level of neurones. If this activity indeed involves quantum effects, we should investigate what the quantum-level analogue of a computer actually is. There is some theoretical work on this question: in particular, it has been shown that there are certain mathematical problems that are solved more quickly by a quantum computer (as defined by David Deutsch) than by a conventional computer—although the problems for which this is so far known to be true constitute a very special class. Quantum computers are potentially more efficient than classical computers for certain types of problem. It may be that this efficiency is an important feature of a quantum-computational behaviour in microtubules. My own purposes would require something more: a situation in which the quantum system performs actions that cannot be explained in computational terms at all.

There are indeed mathematical problems of this nature: problems that cannot be solved using ordinary computational procedures. Yet the insights that human beings have seem to lie beyond computation too, so that if we believe that what we do with our brains is rooted in physical activity, which I *do* believe, the implication is that this physical activity contains some non-computable elements. This is viewpoint C in the following table of alternatives.

Common computers, impressive and wonderful though they may be, are not sufficient to do what brains are capable of. We need something more. I believe that we need a new theoretical grasp of the relationship between quantum-level and classical physics. When we move from one level to the other, existing theory looks rather muddled. We have these things called 'measurements', which give rise to quantum *probabilities* as I discussed earlier. Somehow we change the quantum-mechanical rules when we shift from the small to the large scale when a 'measurement' is performed. The new theory that is needed to make sense of quantum theory will also be required to explain how non-computational activity can apparently arise in some physical systems—in particular, in physical brains.

Why do I say that there ought to be non-computational activity in brain action? The argument comes down to the work of Gödel, the significance of which was referred to by Freeman Dyson in his contribution. I think Gödel's

---

A: All thinking is computation and the mere carrying out of appropriate computations will evoke feelings of conscious awareness.

B: It is the brain's physical action that evokes awareness; and any physical action can in principle be simulated computationally. But computational simulation by itself cannot evoke awareness.

C: Appropriate physical action of the brain evokes awareness, but this physical action cannot even be properly simulated computationally.

D: Awareness cannot be explained in physical, computational, or any other scientific terms.

---

theorem is indeed one of the greatest achievements of the twentieth century. One implication of the Gödel argument is that mathematical belief cannot be encapsulated by any knowable algorithm. The 'knowable' part is the important thing here. What the argument of Gödel tells us is that we cannot know any algorithm A that we are to believe underlies our mathematical beliefs. If we know and believe A, then we must believe additionally that A is consistent. It must not give a nonsense result such as $1 = 2$, for example. If we believe that A is consistent, we must believe its Gödel statement, which we shall call $G(A)$, where $G(A)$ codes the consistency of A. $G(A)$, however, lies beyond the scope of A. The system A cannot encapsulate $G(A)$, so long as A actually is consistent. It follows that there cannot be any A that can be known to us to capture the full extent of our mathematical beliefs.

It seems to me appropriate to conclude that we do not use any definite algorithm at all—but the Gödel argument does not quite show this. What the Gödel argument does not discuss is whether we are acting in accordance with an unconscious unknowable algorithm—or at least an algorithm whose *soundness* is not knowable. Mathematicians, you might well think, should know what they are doing. But if they are in reality the servants of some hidden algorithm operating in their heads then they cannot know what this algorithm is. That they are acting in this unknowable way is a possible but, I think, an unlikely proposal. It opens up a much longer discussion. Up to this point I have employed fairly rigorous mathematics. To go beyond this, it seems that we must turn to other kinds of argument. Are we, as mathematicians,

Usual form of Gödel argument

Formal system $F \begin{cases} \text{axioms} \\ \text{rules of procedure} \end{cases}$

can form Gödel proposition G(F).

If F is **consistent** (and broad enough for arithmetic), then

G(F) is $\begin{cases} \text{not derivable within } F \\ \text{TRUE} \end{cases}$

N.B. F consistent $\Leftrightarrow$ 0 = 1 cannot be proved in F.

Belief   If we believe F sound
                then we believe F consistent
        therefore we believe G(F)   TRUE

Therefore belief is non-computational.

really acting in accordance with an unconscious unknowable algorithm? One inference from such a proposal would be that the reasons we offer for believing our mathematical results are not the true reasons for such belief. Mathematics would depend upon some unknown calculational activity of which we were never aware.

Although this is possible, it seems to me unlikely as the real explanation for mathematical conviction. We have to ask ourselves how this unconscious unknowable algorithm, of value only for doing sophisticated mathematics, could have arisen by a process of natural selection. An alternative argument, which is a kind of composite of the strong artificial intelligence view and a religious one, has been brought to my attention. Its proponents would accept that while the brain may be a kind of computer, it has been programmed by 'the best programmer in the business'. *We* cannot know this mysterious algorithm, but *God* does and it was *he* who instilled it within us, rather than its having arisen by a process of natural selection.

My personal belief is that our mathematical understanding is a product of natural selection. But I would argue that this quality of 'understanding' cannot be an algorithmic thing at all. If it were algorithmic it would have to provide a formal mathematical system that was specific to doing sophisticated mathematics. Such things are very far from ordinary human experience. If the ability to perform advanced mathematics had had a strong selective advantage in the past, there might have been a straightforward case for the contrary belief; but there is no indication of any such selective advantage. There is not even an advantage now! This makes it very hard to see how an algorithm specific to sophisticated mathematics could ever be selected for in evolution. We need the selective processes to have acted in a context in which the very quality of 'understanding' itself—necessarily a non-algorithmic thing—can have had a fundamental selective advantage.

I claim that the Gödel argument demonstrates that whatever understanding is, it is indeed *not* a computational thing. This allows that natural selection could operate for this general non-computational quality—a quality which could be applied to a whole range of problems and not simply to mathematics.

If mentality is a function of brain action, and we accept that brain action is subject to the same laws of physics as everything else, those physical laws must allow room for a non-computational action. The most promising place we should explore in order to locate this action is the region between the quantum and classical levels of physics. This is an area that is ill understood at present, but I believe it is crucial to the issue of non-computational physical activity.

REFERENCES

Hameroff, S. R. (1987). *Ultimate computing*. North Holland, Amsterdam.
Penrose, R. (1989). *The Emperor's new mind*. Oxford University Press.
Penrose, R. (1991). On the cohomology of impossible figures. *Topologie Structurale*, **17**, 11–16.
Penrose, R. (1994). *Shadows of the mind*. Oxford University Press.
Redhead, M. L. G. (1987). *Incompleteness, nonlocality and realism*. Clarendon Press, Oxford.

CHAPTER THREE

Randomness in arithmetic and the decline and fall of reductionism in pure mathematics

GREGORY J. CHAITIN

Hilbert's idea is the culmination of two thousand years of mathematical tradition going back to Euclid's axiomatic treatment of geometry, going back to Leibniz's dream of a symbolic logic and Russell and Whitehead's monumental *Principia Mathematica*. Hilbert's dream was to clarify once and for all the methods of mathematical reasoning. Hilbert wanted to formulate a formal axiomatic system which would encompass all mathematics.

Hilbert emphasized a number of key properties that such a formal axiomatic system should have. It's like a computer programming language. It's a precise statement about the methods of reasoning, the postulates, and the methods of inference that we accept as mathematicians. Furthermore Hilbert stipulated that the formal axiomatic system encompassing all mathematics that he wanted to construct should be 'consistent' and it should be 'complete'.

'Consistent' means that you shouldn't be able to prove an assertion and the contrary of the assertion. You shouldn't be able to prove A and not A. That would be very embarrassing.

'Complete' means that if you make a meaningful assertion you should be able to settle it one way or the other. It means that either A or not A should be a theorem, should be provable from the axioms using the rules of inference in the formal axiomatic system.

Consider a meaningful assertion A and its contrary not A. Exactly one of the two should be provable if the formal axiomatic system is consistent and complete.

A formal axiomatic system is like a programming language. There's an alphabet and rules of grammar; in other words, a formal syntax. It's a kind of thing that we are familiar with now. Look back at Russell and Whitehead's three enormous volumes full of symbols and you will feel that you are looking at a large computer program in some incomprehensible programming language.

Now there's a very surprising fact. 'Consistent and complete' means only truth and all the truth. They seem like reasonable requirements. There's a funny consequence, though, having to do with something called the decision problem. In German it's called the *Entscheidungsproblem*. Hilbert ascribed a great deal of importance to the decision problem.

Solving the decision problem for a formal axiomatic system is giving an algorithm that enables you to decide whether any given meaningful assertion is a theorem or not. A solution of the decision problem is called a decision procedure.

This sounds weird. The formal axiomatic system that Hilbert wanted to construct would have included all mathematics: elementary arithmetic, calculus, algebra, everything. If there's a decision procedure, then mathematicians are out of work. This algorithm, this mechanical procedure, can check whether something is a theorem or not, can check whether it's true or not. So to require that there be a decision procedure for this formal axiomatic system sounds as if we are asking for a lot.

However, it's very easy to see that if it's consistent and it's complete that implies that there must be a decision procedure. This is how you do it. You have a formal language with a finite alphabet and a grammar. And Hilbert emphasized that the whole point of a formal axiomatic system is that there must be a mechanical procedure for checking whether a purported proof is correct or not, whether it obeys the rules or not. That's the notion that mathematical truth should be objective so that everyone can agree whether a proof follows the rules or not.

So if that's the case you run through all possible proofs in size order, and look at all sequences of symbols from the alphabet one character long, two, three, four, a thousand, a thousand and one . . . a hundred thousand characters long. You apply the mechanical procedure, which is the essence of the formal axiomatic system, to check whether each proof is valid. Most of the time, of course, it'll be nonsense, it'll be ungrammatical. But you'll eventually find every possible proof. It's like a million monkeys typing away. You'll find every possible proof, though only in principle of course. The number grows exponentially and this is something that you couldn't do in practice. You'd never get to proofs that are one page long.

But in principle you could run through all possible proofs, check which ones are valid, see what they prove, and that way you can systematically find all theorems. In other words, there is an algorithm, a mechanical procedure, for generating one by one every theorem that can be demonstrated in a formal axiomatic system. So if, for every meaningful assertion within the system, either the assertion is a theorem or its contrary is a theorem, only one of them, then you get a decision procedure. To see whether an assertion is a theorem or not you just run through all possible proofs until you find the assertion coming out as a theorem or you prove the contrary assertion.

So it seems that Hilbert actually believed that he was going to solve, once and for all, all mathematical problems. It sounds amazing, but apparently he did. He believed that he would be able to set down a consistent and complete formal axiomatic system for all mathematics and from it obtain a decision procedure for all mathematics. This is just following the formal, axiomatic tradition in mathematics.

But I'm sure he didn't think that it would be a practical decision procedure. The one I've outlined would work only in principle. It's exponentially slow: terribly slow! Totally impractical. But the idea was that if all mathematicians could agree whether a proof is correct and be consistent and complete, in principle that would give a decision procedure for automatically solving any mathematical problem. This was Hilbert's magnificent dream, and it was to be the culmination of Euclid and Leibniz, and Boole and Peano, and Russell and Whitehead.

Of course the only problem with this inspiring project is that it turned out to be impossible!

GÖDEL, TURING, AND CANTOR'S DIAGONAL ARGUMENT

Hilbert is indeed inspiring. His famous lecture in 1900 is a call to arms to mathematicians to solve a list of twenty-three difficult problems. As a young mathematician you read that list and realize that Hilbert is saying that there is no limit to what mathematicians can do. We can solve a problem if we are clever enough and work at it long enough. He didn't believe that in principle there was any limit to what mathematics could achieve.

I think this is very inspiring. So did John von Neumann. When he was a young man he tried to carry through Hilbert's ambitious programme, because Hilbert couldn't quite get it all to work; Hilbert in fact started off just with elementary number theory, 1, 2, 3, 4, 5, . . ., not even with real numbers at first.

And then in 1931 to everyone's great surprise (including von Neumann's), Gödel showed that it was impossible, that it couldn't be done.

This was the opposite of what everyone had expected. Von Neumann said it never occurred to him that Hilbert's programme couldn't be carried out. Von Neumann admired Gödel enormously, and helped him to get a permanent position at the Institute for Advanced Study.

What Gödel showed was the following. Suppose that you have a formal axiomatic system dealing with elementary number theory, with 1, 2, 3, 4, 5 and addition and multiplication. And we'll assume that it's consistent, which is a minimum requirement—if you can prove false results it's really pretty bad. What Gödel showed was that if you assume that it's consistent, then you can show that it's incomplete. The proof is very clever and involves self-reference. Gödel was able to construct an assertion about the whole numbers that says of itself that it's unprovable. This was a tremendous shock. Gödel has to be admired for his intellectual imagination; everyone else thought that Hilbert was right. I think, however, that Turing's 1936 approach is better.

Gödel's 1931 proof is very ingenious: a real *tour de force*. I have to confess that when I was a kid trying to understand it, I could read it and follow it step by step but somehow I couldn't ever really feel that I was grasping it. Now Turing had a completely different approach.

Turing's approach I think it's fair to say is in some ways more fundamental. In fact, Turing did more than Gödel. Turing not only got as a corollary Gödel's result, he showed that there could be no decision procedure.

You see, if you assume that you have a formal axiomatic system for arithmetic and it's consistent, from Gödel you know that it can't be complete, but there still might be a decision procedure. There still might be a mechanical procedure which would enable you to decide if a given assertion is true or not. That was left open by Gödel, but Turing settled it. The fact that there cannot be a decision procedure is more fundamental and you get incompleteness as a corollary.

How did Turing do it? I want to tell you how he did it because that's the springboard for my own work. The way he did it has to do with something called the halting problem. In fact if you go back to Turing's 1936 paper you will not find the words 'halting problem'. But the idea is certainly there.

People also forget that Turing was talking about 'computable numbers'. The title of his paper is 'On computable numbers, with an application to the *Entscheidungsproblem*'. Everyone remembers that the halting problem is unsolvable and that comes from that paper, but not as many people remember that Turing was talking about computable real numbers. My work deals with computable and dramatically uncomputable real numbers.

Turing's argument is really what destroys Hilbert's dream, and it's a simple argument. It's just Cantor's diagonal procedure applied to the computable real numbers. That's the whole idea in a nutshell, and it's enough to show that

Hilbert's dream, the culmination of two thousand years of what mathematicians thought mathematics was about, is wrong. So Turing's work is tremendously deep.

What is Turing's argument? A real number, say, 3.1415926 . . ., is a length measured with arbitrary precision, with an infinite number of digits. And a computable real number, said Turing, is one for which there is a computer program or algorithm for calculating the digits one by one. For example, there are programs for π, and there are algorithms for solutions of algebraic equations with integer coefficients. In fact most of the numbers that you actually find in analysis are computable. However they're the exception; in terms of set theory, this is because the computable reals are denumerable and the reals are nondenumerable (you don't have to know what that means). That's the essence of Turing's idea.

The idea is this. You list all possible computer programs. At that time there were no computer programs, and Turing had to invent the Turing machine, which was a tremendous step forward. But now you just say, imagine writing a list with every possible computer program.

If you consider computer programs to be in binary, then it's natural to think of a computer program as a natural number. And next to each computer program, the first one, the second one, the third one, write out the real number that it computes if it computes a real (it may not). But if it prints out an infinite number of digits, write them out. So perhaps it's 3.1415926, and here you have another and another and another.

So you make this list. Perhaps some of these programs don't print out an infinite number of digits, because they're programs that halt or that have an error in them and explode. But then there will just be a blank line in the list.

It's not really important—let's forget about this possibility.

Following Cantor, Turing says go down the diagonal and look at the first digit of the first number, the second digit of the second, the third . . .

$$p_1 \text{---} .\underline{d_{11}}d_{12}d_{13}d_{14}d_{15}d_{16} \cdots$$
$$p_2 \text{---} .d_{21}\underline{d_{22}}d_{23}d_{24}d_{25}d_{26} \cdots$$
$$p_3 \text{---} .d_{31}d_{32}\underline{d_{33}}d_{34}d_{35}d_{36} \cdots$$
$$p_4 \text{---} .d_{41}d_{42}d_{43}\underline{d_{44}}d_{45}d_{46} \cdots$$
$$p_5$$
$$p_6 \text{---} .d_{61}d_{62}d_{63}d_{64}d_{65}\underline{d_{66}} \cdots$$
$$\vdots$$

Actually it's the digits after the decimal point. So it's the first digit after the decimal point of the first number, the second digit after the decimal point of the second, the third digit of the third number, the fourth digit of the fourth, the fifth digit of the fifth. And it really doesn't matter if the fifth program doesn't put out a fifth digit.

What you do is you change these digits: make them different. Change every digit on the diagonal. Put these changed digits together into a new number with a decimal point in front, a new real number. That's Cantor's diagonal procedure. So you have a digit which you choose to be different from the first digit of the first number, the second digit of the second, the third of the third and you put these together into one number.

$$p_1 \!-\! .\underline{d_{11}}d_{12}d_{13}d_{14}d_{15}d_{16}\cdots$$
$$p_2 \!-\! .d_{21}\underline{d_{22}}d_{23}d_{24}d_{25}d_{26}\cdots$$
$$p_3 \!-\! .d_{31}d_{32}\underline{d_{33}}d_{34}d_{35}d_{36}\cdots$$
$$p_4 \!-\! .d_{41}d_{42}d_{43}\underline{d_{44}}d_{45}d_{46}\cdots$$
$$p_5$$
$$p_6 \!-\! .d_{61}d_{62}d_{63}d_{64}d_{65}\underline{d_{66}}\cdots$$
$$\vdots$$
$$.\neq d_{11}\neq d_{22}\neq d_{33}\neq d_{44}\neq d_{55}\neq d_{66}\cdots$$

This new number cannot be in the list because of the way it was constructed. Therefore it's an uncomputable real number. How does Turing go on from here to the halting problem? Well, just ask yourself *why* you can't compute it. I've explained how to get this number and it looks as though you could almost do it. To compute the *n*th digit of this number, you get the *n*th computer program (you can certainly do that) and then you start it running until it puts out an *n*th digit, and at that point you change it. That sounds easy.

The problem is, what happens if the *n*th computer program never puts out an *n*th digit, and you sit there waiting? And that's the halting problem—you cannot decide whether the *n*th computer program will ever put out an *n*th digit. This is how Turing got the unsolvability of the halting problem. Because if you could solve the halting problem, then you could decide if the *n*th computer program ever puts out an *n*th digit. And if you could do that then you could actually carry out Cantor's diagonal procedure and compute a real number which has to differ from any computable real. That's Turing's original argument.

Why does this explode Hilbert's dream? What has Turing proved? That there is no algorithm, no mechanical procedure, that will decide if the *n*th computer program ever puts out an *n*th digit. Thus there can be no algorithm that will decide if a computer program ever halts (finding the *n*th digit put out by the *n*th program is a special case). Well, what Hilbert wanted was a formal axiomatic system from which all mathematical truth should follow, only mathematical truth, and all mathematical truth. If Hilbert could do that, it would give us a mechanical procedure to decide if a computer program will ever halt. Why?

You just run through all possible proofs until either you find a proof that the program halts or you find a proof that it never halts. So if Hilbert's dream of a finite set of axioms from which all mathematical truth should follow were possible, then by running through all possible proofs and checking which ones were correct, you would be able to decide whether any computer program halts. In principle you could. But you *can't* by Turing's very simple argument, which is just Cantor's diagonal argument applied to the computable reals. That's how simple it is!

Gödel's proof is ingenious and difficult. Turing's argument is so fundamental, so deep, that everything seems natural and inevitable. But he was of course building on Gödel's work.

THE HALTING PROBABILITY AND ALGORITHMIC RANDOMNESS

The reason why I discussed Turing and computable reals is that I'm going to use a different procedure to construct an uncomputable real, a much more uncomputable real than Turing does. And that's how we're going to get into much worse trouble.

How do I get a much more uncomputable real? (And I'll have to tell you how uncomputable it is.) Well, not with Cantor's diagonal argument. I get this number, which I like to call Ω, like this:

$$\Omega = \sum_{p\,\text{halts}} 2^{-|p|}.$$

This is the halting probability. It's sort of a mathematical pun. Turing's fundamental result is that the halting problem is unsolvable: there is no algorithm that'll settle the halting problem. My fundamental result is that the halting probability is algorithmically irreducible or algorithmically random.

Instead of looking at individual programs and asking whether they halt, you put all computer programs together in a bag. If you generate a computer program at random by tossing a coin for each bit of the program, what is the chance that the program will halt? You're thinking of programs as bit strings, and you generate each bit by an independent toss of a fair coin, so if a program is N bits long, then the probability that you get that particular program is 2^{-N}. Any program p that halts contributes $2^{-|p|}$, two to the minus its size in bits, the number of bits in it, to this halting probability.

By the way, there's a technical detail which is very important and didn't work in the early version of algorithmic information theory. You couldn't in fact use

$$\Omega = \sum_{p\,\text{halts}} 2^{-|p|}.$$

It would give infinity. The technical detail is that no extension of a valid program is a valid program. The sum I have just given,

$$\sum_{p\,\text{halts}} 2^{-|p|},$$

then turns out to be between zero and one. Otherwise it turns out to be infinity. It took only ten years until I got it right. The original 1960s version of algorithmic information theory is wrong. One of the reasons it's wrong is that you can't even define this number

$$\Omega = \sum_{p\,\text{halts}} 2^{-|p|}.$$

In 1974 I redid algorithmic information theory with 'self-delimiting' programs and then I discovered the halting probability Ω.

This, then, is a probability between zero and one,

$$0 < \Omega = \sum_{p\,\text{halts}} 2^{-|p|} < 1,$$

like all probabilities. The idea is that you generate each bit of a program by tossing a coin; then ask, What is the probability that it halts? The number Ω, the halting probability, is not only an uncomputable real—Turing already knew how to do that. It is uncomputable in the worst possible way. Let me give you some clues to show how uncomputable it is.

Well, for one thing it's algorithmically incompressible. If you want to get the first n bits of Ω out of a computer program, if you want a computer program that will print out the first n bits of Ω and then halt, that computer program has to be n bits long. Essentially you're only printing out constants that are in the program. You cannot squeeze the first n bits of Ω. This

$$0 < \Omega = \sum_{p\,\text{halts}} 2^{-|p|} < 1$$

is a real number; you could write it in binary. And if you want to get out the first n bits from a computer program, essentially you just have to put them in. The program has to be n bits long. That's irreducible algorithmic information. There is no concise description.

Now that's an abstract way of saying things.

Someone may ask why Ω should be a probability. What if the two one-bit programs both halt? What if the two one-bit programs both halt and then some other program halts? Then Ω is greater than one and not a probability.

The answer is that no extension of a valid program is a valid program; no other programs can halt. The two one-bit programs would be all the programs there are. That's why the number

$$0 < \Omega = \sum_{p \text{ halts}} 2^{-|p|} < 1$$

can't be defined if you think of programs in the normal way.

Let me give a more concrete example of how random Ω is. Emile Borel at the turn of this century was one of the founders of probability theory, and he talked about something he called a normal number.

What is a normal real number? People have calculated π to a billion digits, perhaps two billion. One of the reasons for doing this is that it's like climbing a mountain and having the world record; another is the question of whether each digit occurs the same number of times. It looks as though the digits 0 through 9 each occur 10 per cent of the time in the decimal expansion of π. It looks that way, but nobody can prove it. I think the same is true for $\sqrt{2}$, although that's not as popular a number to ask this about.

Let me describe some work that Borel did around the turn of the century when he was pioneering modern probability theory. Pick a real number in the unit interval, a real number with a decimal point in front, with no integer part. If you pick a real number in the unit interval, Borel showed that with probability one it's going to be 'normal'. 'Normal' means that when you write it in decimal each digit will occur in the limit exactly 10 per cent of the time, and this will also happen in any other base. For example, in binary 0 and 1 will each occur in the limit exactly 50 per cent of the time. Similarly with blocks of digits. This was called an absolutely normal real number by Borel, and he showed that with probability one if you pick a real number at random between zero and one it's going to have this property. There's only one problem. He didn't know whether π was normal, he didn't know whether $\sqrt{2}$ was normal. In fact, he couldn't exhibit a single individual example of a normal real number.

The first example of a normal real number was discovered by a friend of Alan Turing's at Cambridge called David Champernowne, who, at the time of writing, is still alive and who is a well-known economist. Turing was impressed with him—I think he called him 'Champ'—because Champ had published this in a paper as an undergraduate. This number is known as Champernowne's number. Let me show you Champernowne's number:

$$0.01234567891011121314 \ldots 99100101 \ldots$$

It goes like this. You write down a decimal point, then you write 0, 1, 2, 3, 4, 5, 6, 7, 8, 9, then 10, 11, 12, 13, 14 up to 99, then 100, 101. And you keep going in this funny way. This is called Champernowne's number and Champernowne showed that it's normal in base ten, but only in base ten. Nobody knows if it's normal in other bases; I think that's still open. In base ten, though, not only will the digits 0 through 9 occur exactly 10 per cent of the time in the limit, but each possible block of two digits will occur exactly 1 per cent of the time in the limit, each block of three digits will occur exactly 0.1 per cent of the time in the limit, etc. That's called being normal in base ten. But nobody knows what happens in other bases.

The reason why I'm saying all this is because it follows from the fact that the halting probability Ω is algorithmically irreducible information that

$$0 < \Omega = \sum_{p \text{ halts}} 2^{-|p|} < 1$$

is normal in any base. That's easy to prove using ideas about coding and compressing information that go back to Shannon. So here we finally have an example of an absolutely normal number. I don't know how natural you think it is, but it is a specific real number that comes up and is normal in the most demanding sense that Borel could think of. Champernowne's number couldn't quite do that.

The number Ω is in fact random in many more senses. I would put it this way. It cannot be distinguished from the result of independent tosses of a fair coin. In fact this number Ω shows that you have total randomness and chaos and unpredictability and lack of structure in pure mathematics! In the same way that all it took for Turing to destroy Hilbert's dream was the diagonal argument, you just write down this expression:

$$0 < \Omega = \sum_{p \text{ halts}} 2^{-|p|} < 1,$$

and this shows that there are regions of pure mathematics where reasoning is totally useless, where you're up against an impenetrable wall. This is all it takes. It's just this halting probability.

Why do I say this? Well, let's say you want to use axioms to prove what the bits of this number Ω are. I've already said that it's uncomputable—like the number that Turing constructs using Cantor's diagonal argument. So we know there is no algorithm that will compute digit by digit or bit by bit this number Ω. But let's try to prove what individual bits are using a formal axiomatic system. What happens?

The situation is very, very bad. It's like this. Suppose you have a formal axiomatic system which is N bits of formal axiomatic system (I'll explain what this means more precisely later). It turns out that with a formal axiomatic system of complexity N, that is, N bits in size, you can prove what the positions and values are of at most $(N+c)$ bits of Ω.

Now what do I mean by formal axiomatic system N bits in size? Well, remember that the essence of a formal axiomatic system is a mechanical procedure for checking whether a formal proof follows the rules or not. It's a computer program. Of course in Hilbert's days there were no computer programs, but after Turing invented Turing machines you could finally specify the notion of a computer program exactly, and now we're very familiar with it.

So the proof-checking algorithm which is the essence of any formal axiomatic system in Hilbert's sense is a computer program. Just see how many bits long this computer program is. (A technical point: actually, it's best to think of the complexity of a formal axiomatic system as the size in bits of the computer program that enumerates the set of all theorems.) That's essentially how many bits it takes to specify the rules of the game, the axioms and postulates and the rules of inference. If that's N bits, then you may be able to prove, say, that the first bit of Ω in binary is 0, that the second bit is 1, that the third bit is 0, and then there might be a gap, and you might be able to prove that the thousandth bit is 1. But you're going to be able to settle only N cases if your formal axiomatic system is an N-bit formal axiomatic system.

Let me try to explain better what this means. It means that you can get out only as much as you put in. If you want to prove whether an individual bit in a specific place in the binary expansion of the real number Ω is a 0 or a 1, essentially the only way to prove that is to take it as a hypothesis, as an axiom, as a postulate. It's irreducible mathematical information. That's the key phrase that really gives the whole idea.

So what have we got? We have a rather simple mathematical object that completely escapes us. Ω's bits have no structure. There is no pattern, there is no structure that we as mathematicians can comprehend. If you're interested in proving what individual bits of this number at specific places are, whether they're 0 or 1, reasoning is completely useless. Here mathematical reasoning is irrelevant and can get nowhere. As I said before, the only way a formal axiomatic system can get out these results is essentially just to put them in as assumptions, which means you're not using reasoning. After all, anything can be demonstrated by taking it as a postulate that you add to your set of axioms. So this is a worst possible case—this is irreducible mathematical information. Here is a case where there is no structure, there are no correlations, there is no pattern that we can perceive.

RANDOMNESS IN ARITHMETIC

What does this have to do with randomness in arithmetic? Now we're going back to Gödel: I skipped over him rather quickly, and now let's go back.

Turing says that you cannot use proofs to decide whether a program will halt. You can't always prove that a program will halt or not. That's how he destroys Hilbert's dream of a universal mathematics. I get us into more trouble by looking at a different kind of question, namely, can you prove that the fifth bit of this particular real number

$$0 < \Omega = \sum_{p\ \text{halts}} 2^{-|p|} < 1$$

is a 0 or a 1, or that the eighth bit is a 0 or a 1? But these are strange-looking questions. Who had ever heard of the halting problem in 1936? These are not the kind of things that mathematicians normally worry about. We're getting into trouble, but with questions rather far removed from normal mathematics.

Even though you can't have a formal axiomatic system that can always prove whether a program halts or not, it might be good for everything else and then you could have an *amended* version of Hilbert's dream. And the same with the halting probability Ω. If the halting problem looks a little bizarre, and it certainly did in 1936, well, Ω is brand new and certainly looks bizarre. Who ever heard of a halting probability? It's not the kind of thing that mathematicians normally do. So what do I care about all these incompleteness results?

Well, Gödel had already faced this problem with his assertion which is true but unprovable. It's an assertion which says of itself that it's unprovable. That kind of thing also never comes up in real mathematics. One of the key elements in Gödel's proof is that he managed to construct an *arithmetical* assertion which says of itself that it's unprovable. It was getting this self-referential assertion to be in elementary number theory which took so much cleverness.

There has since been a lot of work building on Gödel's work, showing that problems involving computations are equivalent to arithmetical problems involving whole numbers. A number of names come to mind. Julia Robinson, Hilary Putnam, and Martin Davis did some of the important work, and then a key result was found in 1970 by Yuri Matijasevič. He constructed a diophantine equation, which is an algebraic equation involving only whole numbers, with a lot of variables. One of the variables, K, is distinguished as a parameter. It's a polynomial equation with integer coefficients and all the unknowns have to be whole numbers. As I said, one of the unknowns is a parameter. Matijasevič's equation has a solution for a particular value of the parameter K if and only if the Kth computer program halts.

In the year 1900 Hilbert had asked for an algorithm that would decide whether a diophantine equation has a solution. This was Hilbert's tenth problem. It was tenth in his famous list of twenty-three problems. What

Matijasevič showed in 1970 was that this is equivalent to deciding whether an arbitrary computer program halts. So Turing's halting problem is exactly as hard as Hilbert's tenth problem. It's exactly as hard to decide whether an arbitrary program will halt as to decide whether an arbitrary algebraic equation in whole numbers has a solution. Therefore there is no algorithm for doing that and Hilbert's tenth problem cannot be solved: that was Matijasevič's 1970 result.

Matijasevič has gone on working in this area. In particular there is a piece of work he did in collaboration with James Jones in 1984. I can use it to follow in Gödel's footsteps, to follow Gödel's example. You see, I've shown that there's complete randomness, no pattern, lack of structure, and that reasoning is completely useless, if you're interested in the individual bits of this number

$$0 < \Omega = \sum_{p \text{ halts}} 2^{-|p|} < 1.$$

Following Gödel, let's convert this into something in elementary number theory. Because if you can get into all this trouble in elementary number theory, that's the bedrock. Elementary number theory, 1, 2, 3, 4, 5, addition and multiplication: that goes back to the ancient Greeks and it's the most solid part of all of mathematics. In set theory you're dealing with strange objects like large cardinals, but here you're not even dealing with derivatives or integrals or measure, only with whole numbers. And using the 1984 results of Jones and Matijasevič I can indeed dress up Ω arithmetically and get randomness in elementary number theory.

What I get is an exponential diophantine equation with a parameter. 'Exponential diophantine equation' just means that you allow variables in the exponents. In contrast, what Matijasevič used to show that Hilbert's tenth problem is unsolvable is just a polynomial diophantine equation, which means that the exponents are always natural number constants. I have to allow X^Y. It's not known yet whether I actually need to do this. It might be the case that I can manage with a polynomial diophantine equation: it's still an open question. But for now, what I have is an exponential diophantine equation with seventeen thousand variables. This equation is two hundred pages long and again one variable is the parameter.

This is an equation where every constant is a whole number, a natural number, and all the variables are also natural numbers, that is, positive integers (actually *non-negative* integers). One of the variables is a parameter, and you change the value of this parameter—take it to be 1, 2, 3, 4, 5. Then you ask, does the equation have a finite or infinite number of solutions? My equation is constructed so that it has a finite number of solutions if a particular individual bit of Ω is a 0, and it has an infinite number of solutions if that bit is a 1. So, deciding whether my exponential diophantine equation in each individual case has a finite or infinite number of solutions is exactly the same as determining what an individual bit of the halting probability Ω is. And this is completely intractable because Ω is irreducible mathematical information.

Let me emphasize the difference between this and Matijasevič's work on Hilbert's tenth problem. Matijasevič showed that there is a polynomial diophantine equation with a parameter with the following property: you vary the parameter and ask, does the equation have a solution? That turns out to be equivalent to Turing's halting problem, and therefore escapes the power of mathematical reasoning, of formal axiomatic reasoning.

How does this differ from what I do? I use an exponential diophantine equation, which means I allow variables in the exponent. Matijasevič allows only constant exponents. The big difference is that Hilbert asked for an algorithm to decide if a diophantine equation has a solution. The question I have to ask to get randomness in elementary number theory, in the arithmetic of the natural numbers, is slightly more sophisticated. Instead of asking whether there is a solution, I ask whether there are a finite or infinite number of solutions—a more abstract question. This difference is necessary.

My two-hundred page equation is constructed so that it has a finite or infinite number of solutions depending on whether a particular bit of the halting probability is a 0 or a 1. As you vary the parameter, you get each individual bit of Ω. Matijasevič's equation is constructed so that it has a solution if and only if a particular program ever halts. As you vary the parameter, you get each individual computer program.

Thus, even in arithmetic you can find Ω's absolute lack of structure, Ω's randomness and irreducible mathematical information. Reasoning is completely powerless in those areas of arithmetic. My equation shows that this is so. As I said before, to get this equation I use ideas that start in Gödel's original 1931 paper. But it was Jones and Matijasevič's 1984 paper that finally gave me the tool that I needed.

So that's why I say that there is randomness in elementary number theory, in the arithmetic of the natural numbers. This is an impenetrable stone wall, it's a worst case. From Gödel we knew that we couldn't get a formal axiomatic system to be complete. We knew we were in trouble, and Turing showed us how basic it was, but Ω is an extreme case where reasoning fails completely.

I won't go into the details, but let me talk in vague information-theoretic terms. Matijasevič's equation gives you N arithmetical questions with yes/no answers that turn out to be only log N bits of algorithmic information. My equation gives N arithmetical questions with yes/no answers that are irreducible, incompressible mathematical information.

EXPERIMENTAL MATHEMATICS

Let me add a little about what this all means.

First of all, the connection with physics. There was a big controversy when quantum mechanics was developed, because quantum theory is nondeterministic. Einstein didn't like that. He said, 'God doesn't play dice!' But with chaos and nonlinear dynamics we've now realized that even in classical physics we get randomness and unpredictability. My work is in the same spirit. It shows that pure mathematics, in fact even elementary number theory, the arithmetic of the natural numbers, is in the same boat. We get randomness there too. So, as a newspaper headline would put it, God not only plays dice in quantum mechanics and in classical physics, but even in pure mathematics, even in elementary number theory. So if a new paradigm is emerging, randomness is at the heart of it. Randomness is also at the heart of quantum field theory, as virtual particles and Feynman path integrals (sums over all histories) show very clearly. So my work fits in with a lot of work in physics, which is why I often get invited to talk at physics meetings.

However the really important question isn't physics, it's mathematics. I've heard that Gödel wrote a letter to his mother, who stayed in Europe. Gödel and Einstein were friends at the Institute for Advanced Study. You'd see them walking down the street together. Gödel apparently told his mother that even though Einstein's work on physics had really had a tremendous impact on how people did physics, he was disappointed that his work had not had the same effect on mathematicians. It hadn't made a difference to how mathematicians actually carried on their everyday work. So I think that's the key question: how should you really do mathematics?

I'm claiming I have a much stronger incompleteness result. If so, perhaps it'll be clearer whether mathematics should be done the ordinary way. What is the ordinary way of doing mathematics? In spite of the fact that any finite set of axioms is incomplete, how do mathematicians actually work? Well, suppose you have a conjecture that you've been thinking about for a few weeks, and you believe it because you've tested a large number of cases on a computer. Perhaps it's a conjecture about the primes and for two weeks you've tried to prove it. At the end of two weeks you don't say, 'Well obviously the reason I haven't been able to show this is because of Gödel's incompleteness theorem. Let us therefore add it as a new axiom.' But if you took Gödel's incompleteness theorem very seriously this might in fact be the way to proceed. Mathematicians will laugh but physicists actually behave in this way.

Look at the history of physics. We start with Newtonian physics. We cannot get Maxwell's equations from Newtonian physics. It's a new domain of experience—we need new postulates to deal with it. As for special relativity, well, special relativity is almost in Maxwell's equations. But Schrödinger's equation does not come from Newtonian physics and Maxwell's equations. It's a new domain of experience and again we need new axioms. So physicists are used to the idea that when we start experimenting at a smaller scale, or with new phenomena, we may need new principles to understand and explain what's going on.

Now in spite of incompleteness mathematicians don't behave at all like physicists. At a subconscious level they still assume that the small number of principles, of postulates and methods of inference, that they learned early as mathematics students, are enough. In their hearts they believe that if you can't prove a result it's your own fault. That's probably a good attitude to take rather than to blame someone else, but let's look at a question like the Riemann hypothesis. A physicist would say that there is ample experimental evidence for the Riemann hypothesis and would go ahead and take it as a working assumption.

What is the Riemann hypothesis? There are many unsolved questions involving the distribution of the prime numbers that can be settled if you assume the Riemann hypothesis. Using computers people check these conjectures and they work beautifully. They're neat formulas but nobody can prove them. A lot of them follow from the Riemann hypothesis. To a physicist this would be enough: it's useful, it explains a lot of data. Of course a physicist then has to be prepared to say 'Oh, I goofed!' because an experiment can subsequently contradict a theory. This happens very often.

In particles physics you throw up theories all the time and most of them quickly die. But mathematicians don't like to have to back-pedal. But if we play it safe, the problem is that we may be losing out, and I believe we are.

I think it should be obvious where I'm leading. I believe that elementary number theory and the rest of mathematics should be pursued more in the spirit of experimental science, and that we should be willing to adopt new principles. I believe that Euclid's statement that an axiom is a self-evident truth is a big mistake. The Schrödinger equation certainly isn't a self-evident truth! And the Riemann hypothesis isn't self-evident either, but it's very useful.

So I believe that we mathematicians shouldn't ignore incompleteness. It's a safe thing to do but we're losing out on results that we could get. It would be as if physicists said, 'Okay no Schrödinger equation, no Maxwell's equations, we stick with Newton, everything must be deduced from Newton's laws'. (Maxwell even tried it. He had a mechanical model of an electromagnetic field. Fortunately they don't teach that in college!)

I proposed all this twenty years ago when I started getting these information-theoretic incompleteness results. But independently a new school on the philosophy of mathematics is emerging called the 'quasi-empirical' school of thought regarding the foundations of mathematics. There's a book by Tymoczko (1986) that contains a good collection of articles. Casti (1990) has a good chapter on mathematics. The last half of the chapter talks about this quasi-empirical view.

By the way, Lakatos, who was one of the people involved in this new movement, happened to be at Cambridge at that time. He had left Hungary.

The main schools of mathematical philosophy at the beginning of this century were Russell and Whitehead's view that logic was the basis for everything, the formalist school of Hilbert, and an 'intuitionist' constructivist school of Brouwer. Some people think that Hilbert believed that mathematics is a meaningless game played with marks of ink on paper. Not so! He just said that to be absolutely clear and precise what mathematics is all about, we have to specify the rules determining whether a proof is correct so precisely that they become mechanical. Nobody who thought that mathematics is meaningless would have been so energetic and done such important work and been such an inspiring leader.

Originally most mathematicians backed Hilbert. Even after Gödel and even more emphatically Turing showed that Hilbert's dream didn't work, in practice mathematicians carried on as before, in Hilbert's spirit. Brouwer's constructivist attitude was mostly considered a nuisance. As for Russell and Whitehead, they had a lot of problems in getting all of mathematics from logic. If you derive all mathematics from set theory, you discover that it's nice to define the whole numbers in terms of sets (von Neumann worked on this). But then it turns out that there are all kinds of problems with sets. You're not making the natural numbers more solid by basing them on something which is more problematical.

Now everything has gone topsy-turvy: not because of any philosophical argument, not because of Gödel's results, or Turing's results, or my own incompleteness results. It's gone topsy-turvy for a very simple reason: the computer.

The computer, as we all know, has changed the way we do everything. The computer has enormously and vastly increased mathematical experience. It's so easy to do calculations, to test many cases, to run experiments on the computer. The computer has so vastly increased mathematical experience, that in order to cope, mathematicians are forced to proceed in a more pragmatic fashion, more like experimental scientists. This new tendency is often called 'experimental mathematics'. This phrase comes up a lot in the field of chaos, fractals, and nonlinear dynamics.

It's often the case that when doing experiments on the computer, numerical experiments with equations, you see that something happens, and you conjecture a result. Of course it's nice if you can prove it. Especially if the proof is short. I'm not sure that a thousand-page proof helps too much. But if it's a short proof it's certainly better than not having a proof. And if you have several proofs from different viewpoints, that's very good.

But sometimes you can't find a proof and you can't wait for someone else to find a proof, and you've got to carry on as best you can. So now mathematicians sometimes go ahead with working hypotheses on the basis of the results of computer experiments. Of course if it's physicists doing these computer experiments, then it's certainly okay; they've always relied heavily on experiments. But now even mathematicians sometimes operate in this manner. I believe that there's a new journal called the *Journal of Experimental Mathematics*. They ought to have put me on their editorial board, because I've been proposing this for twenty years based on my information-theoretic ideas.

So in the end it wasn't Gödel, it wasn't Turing, and it wasn't my results that are making mathematics go in an experimental mathematics direction, in a quasi-empirical direction. The reason why mathematicians are changing their working habits is the computer. I think it's an excellent joke! (It's also funny that of the three old schools of mathematical philosophy, logicist, formalist, and intuitionist, the most neglected was Brouwer, who had a constructivist attitude years before the computer gave a tremendous impulse to constructivism.)

Of course, the mere fact that everybody is doing something doesn't mean that they ought to be. But I think that the sequence of work that I've outlined does provide some theoretical justification for what everybody's doing anyway, without worrying about the theoretical justification. And I think that the question of how we should actually do mathematics requires *at least* another generation of work.

REFERENCES

Casti, J. (1990). *Searching for certainty*. Morrow, New York.

Chaitin, G. J. (1990). *Algorithmic information theory*, revised 3rd printing. Cambridge University Press.

Chaitin, G. J. (1990). *Information, randomness, and incompleteness*, 2nd edn. World Scientific, Singapore.

Chaitin, G. J. (1992). *Information-theoretic incompleteness*. World Scientific, Singapore.

Chaitin, G. J. (1994). *The limits of mathematics*. Published electronically. To obtain, send e-mail to 'chao-dyn@xyz.lanl.gov' with 'Subject: get 9407010'.

Tymoczko, T. (1986). *New directions in the philosophy of mathematics*. Birkhäuser, Boston.

CHAPTER FOUR

Theories of Everything

JOHN D. BARROW

INTRODUCTION

Despite the topicality of 'Theories of Everything'[1] in the literature of science and its popular chronicles, they are at root a new edition of something very old indeed. If we cast our eye over a range of ancient mythological accounts of the world we soon find that we have before us the first 'theories of everything'. Their authors composed an elaborate story in which there was a place for everything and everything had its place.[2] These were not in any modern sense scientific theories about the world, but tapestries within which the known and the unknown could be interwoven to produce a single picture with a 'meaning', in which the authors could place themselves with a confidence born of their interpretation of the world around them. In time, as more things were discovered and added to the story, so it became increasingly contrived and complicated. Moreover, whilst these accounts aimed at great breadth when assimilating perceived truths about the world into a single coherent whole, they were totally lacking in depth: that is, in the ability to extract more from their story than what was put into it in the first place. Modern scientific theories about the world place great emphasis upon depth—the ability to predict new things and explain phenomena not incorporated in the specification of the theory in the first place. For example, an explanation for the living world that maintained it to have been created ready made just hundreds of years ago, but accompanied by a fossil record with the appearance of billions of years of antiquity, certainly has breadth; but it is shallow. Experience teaches us that it is most efficient to begin with a theory that is narrow but deep and then seek to extend it into a description that is both broad and deep.[1] If we begin with a picture that is broad and shallow, we have insufficient guidance to graduate to a correct description that is broad and deep.

In recent years there has been renewed interest in the possibility of a 'Theory of Everything'.[1] In what follows we shall see what is meant by a 'Theory of Everything' and how, while it may be necessary for our description of the Universe and its contents, it is far from sufficient to complete that understanding. We cannot 'reduce' everything we see to a 'Theory of Everything' of the particle physicists' sort. Other factors must enter to complete the scientific description of the Universe. One of the lessons that will emerge from our account is the extent to which it is dangerous to draw conclusions about 'science', or the 'scientific method' in general, in the course of discussing an issue like reductionism, or the relative merits of religion and science. 'Local sciences', like biology or chemistry, are quite different from astronomy or particle physics. In the local sciences one can gather virtually any data one likes, one can perform any experiment, and (most important of all) one has control over possible sources of bias introduced by the experimental set-up or the process of gathering observations. Experiments can be repeated in different ways. In astronomy this is not the case: we cannot experiment with the Universe: we just have to take what is on offer. What we see is inevitably biased by our existence and how we see it: intrinsically bright objects are invariably over-represented in astronomical surveys. Likewise, in high-energy particle physics there are serious limitations imposed upon our ability to experiment. We cannot achieve the very high energies that are required to unlock many of the secrets of the elementary particle world by direct experiment. The philosophy of science has said a lot about scientific method under the assumption of an ideal environment in which any desired experiment is possible. It has not, to my knowledge, addressed the reality of limited experimental possibility with the same enthusiasm.

ORDER OUT OF CHAOS

Suppose you encounter two sequences of digits. The first has the form

$$\ldots 001001001001001001 \ldots,$$

whilst the second has the form

$$\ldots 010010110101111010010 \ldots$$

Now you are asked if these sequences are random or ordered. Clearly the first appears to be ordered, and you say this because it is possible to 'see' a pattern in it; that is, we can replace the first sequence by a rule that allows us to remember it or convey it to others without simply listing its contents. Thus we will call a sequence non-random if it can be abbreviated by a formula or a rule shorter than itself. If this is so, we say that it is *compressible*.[3] On the other hand, if as appears to be the case for the second sequence (which was generated by tossing a coin), there is no abbreviation or formula that can capture its information content, then we say that it is *incompressible*. If we want to tell our friends about the incompressible sequence then we simply have to list it in full. There is no encapsulation of its information content shorter than itself.

This simple idea allows us to draw some lessons about the scientific search for a Theory of Everything. We might define science to be the search for compressions. We observe the world in all possible ways and gather facts about it; but this is not science. We are not content, like crazed historians, simply to gather up a record of everything that has ever happened. Instead we look for patterns in those facts, compressions of the information on offer, and these patterns we have come to call the laws of Nature. The search for a Theory of Everything is the quest for an ultimate compression of the world. Interestingly, Chaitin's proof of Gödel's incompleteness theorem,[4] using the concepts of complexity and compression, reveals that Gödel's theorem is equivalent to the fact that one cannot prove a sequence to be incompressible. We can never prove a compression to be the ultimate one; there might still be a deeper and simpler unification waiting to be found.

There is a further point that we might raise regarding the quest for a Theory of Everything—if it exists. We might wonder whether such a theory is buried deep (perhaps infinitely deep) in the nature of the Universe or whether it lies rather shallow. One suspects it to lie deep in the structure of things, and so it might appear a most anti-Copernican over-confidence to expect that we would be able to fathom it after just a few hundred years of serious study of the laws of Nature, aided by limited observations of the world by relatively few individuals. There appears to be no good evolutionary reason why our intellectual capabilities need to be so great as to unravel the ultimate laws of Nature, unless those ultimate laws are simply a vast elaboration of very simple principles, like counting or comparing,[5] which are employed in local laws. Of course, the unlikelihood of our success is no reason not to try. We just should not have unrealistic expectations about the chances of success and the magnitude of the task.

LAWS OF NATURE

Our discussion of the compressibility of sequences has taught us that pattern, or symmetry, is equivalent to laws or rules of change.[6] Classical laws of change, like Newton's law of linear momentum conservation, are equivalent to the invariance of some quantity or pattern. These equivalences became known only long after the formulation of the laws of motion governing allowed changes. This strikes a chord with the Platonic tradition which places emphasis upon the unchanging, atemporal aspects of the world, as the key to its fundamental structures. These timeless attributes, or 'forms' as Plato called them, seem to have emerged with the passage of time as the laws of Nature or the invariances and conserved quantities (like energy and momentum) of modern physics.

Since 1973 this focus upon symmetry has taken centre stage in the study of elementary particle physics and the laws governing the fundamental interactions of Nature. Symmetry is now taken as the primary guide into the structure of the elementary particle world, and the laws of change are derived from the requirement that particular symmetries, often of a highly abstract character, be preserved. Such theories are called 'gauge theories'.[7] The most successful theories of the four known forces of Nature—the electromagnetic, weak, strong, and gravitational forces—are all gauge theories. These theories require the existence of the forces they describe in order to preserve the invariances upon which they are based. They are also able to dictate the character of the elementary particles of matter that they govern. In these respects they differ from the classical laws of Newton which, since they governed the motions of all particles, could say nothing about the properties of those particles. The reason for this added dimension is that the elementary particle world that the gauge theories rule, in contrast to the macroscopic world, is populated by collections of *identical* particles. ('Once you've seen one electron you've seen 'em all.')

The use of symmetry in this powerful way enables entire systems of natural laws to be derived from the requirement that a certain abstract pattern be invariant in the Universe. Subsequently, the predictions of this system of laws can be compared with the actual course of Nature. This is the opposite route to that which might have been followed a century ago. Then, the systematic study of events would have led to systems of mathematical equations giving the laws of change; afterwards, the fact that they are equivalent to some global or local invariance might be recognized.

This generation of theories for each of the separate interactions of Nature has motivated the search for a unification of those theories into more comprehensive editions, based upon larger symmetries, within which the smaller symmetries respected by the individual forces of Nature might be accommodated in an interlocking fashion that places some new constraint upon their allowed forms. So far this strategy has resulted in a successful, experimentally tested, unification of the electromagnetic and weak interactions, and a number of purely theoretical proposals for a further unification with the strong interaction ('grand unification'), and ultimately a fourfold unification with the gravitational force to produce a so-called 'Theory of Everything' or 'TOE'.[1] The pattern of unification that has occurred over the last three hundred years is shown in Fig 4.1.

The current favoured candidate for a TOE is a superstring theory, first developed by Michael Green and John Schwartz.[8] Elementary descriptions of its workings can be found elsewhere.[1,9] Suffice it to say that the enormous interest that these theories attracted over the last nine years can be attributed to the fact that they revealed that the requirement of logical self-consistency— suspected of being a rather weak constraint upon a TOE—turned out to be enormously restrictive. At first it was believed that it narrowed the alternatives down to just two possible symmetries underlying the TOE. The situation has subsequently been found to be rather more complicated than first imagined, and superstring theories have been found to require new types of mathematics for their elucidation.

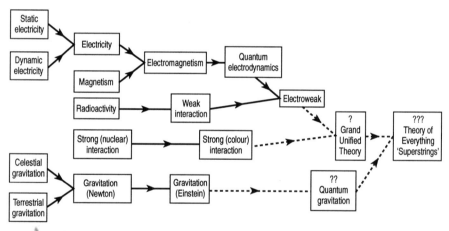

Fig. 4.1. The development of different theories of physics showing how theories of different forces of Nature have become unified and refined. The dotted lines represent unifications that have so far been made only theoretically and have yet to be confirmed by experimental evidence.

The important lesson to be learned from this part of our discussion is that 'Theories of Everything', as currently conceived, are simply attempts to encapsulate all the laws governing the fundamental forces of Nature within a single law of Nature derived from the preservation of a single overarching symmetry. We might add that at present four fundamental forces are known, of which the weakest is gravitation. There might exist other far weaker forces of Nature that are too weak for us to detect (perhaps ever) but whose existence is necessary to fix the logical necessity of that single Theory of Everything.

OUTCOMES AND BROKEN SYMMETRIES

If you were to engage some particle physicists in a conversation about the nature of the world they might soon be regaling you with a story about how simple and symmetrical the world really is if only you would look at things in the right way. But when you return to contemplate the real world you know that it is far from simple. For the psychologist, the economist, the botanist, or the zoologist the world is far from simple. It is a higgledy-piggledy of complex events whose nature owes more to their persistence or stability over time than any mysterious attraction for symmetry or simplicity. So who is right? Is the world really simple, as the physicist said; or is it extremely complicated as everyone else seems to think?[10]

The answer to this question reveals one of the deep subtleties of the Universe's structure. When we look around us we do not observe the laws of Nature; rather, we see the outcomes of those laws. There is a world of difference. Outcomes are much more complicated than the underlying laws because they do not have to respect the symmetries displayed by the laws. By this means it is possible to have a world that displays complicated asymmetrical structures (like ourselves) and yet is governed by very simple, symmetrical laws. Consider the following simple example. Suppose I balance a ball at the apex of a cone. If I were to release the ball then the law of gravitation would determine its subsequent motion. But gravity has no preference for any particular direction in the Universe; it is entirely democratic in that respect. Yet when I release the ball, it will always fall in some particular direction; either because it was given a little push in one direction, or as a result of quantum fluctuations that do not permit an unstable equilibrium state to persist. Thus, in the outcome of the ball falling down, the directional symmetry of the law of gravity is broken. Take another example. You and I are at this moment situated at particular places in the Universe despite the fact that the laws of Nature display no preference for any one place in the Universe over any other. We are both (very complicated) outcomes of the laws of Nature that break their underlying symmetries with respect to positions in space. This teaches us why science is often so difficult. When we observe the world we see only the broken symmetries manifested through the outcomes of the laws of Nature, and from them we must work backwards to unmask the hidden symmetries that characterize the laws behind the appearances.

We can now understand the answers that were given to our first question. The particle physicist works closest to the laws of Nature themselves and so is impressed by their simplicity and symmetry. That is the basis for his confidence in the simplicity of Nature. But the biologist or the meteorologist is occupied with the study of the complex outcomes of the laws, rather than with the laws themselves. As a result she is most impressed by the complexities of Nature rather than by symmetry. This dichotomy is displayed in Fig. 4.2.

The left-hand column represents the development of the Platonic perspective on the world, with its emphasis upon the unchanging elements behind things—laws, conserved quantities, symmetries—whereas the right-hand column, with its stress upon time and change and the concatenation of complex happenings, is the fulfilment of the Aristotelian approach to understanding of the world. Until rather recently, physicists have focused almost exclusively upon the study of the laws rather than the complex outcomes. This is not surprising: the study of the outcomes is a far more difficult problem that requires the existence of powerful interactive computers with good graphics for its full implementation. It is no coincidence that the study of complexity and chaos[11] in that world of outcomes has gone hand in hand with the growing power and availability of low-cost personal computers.

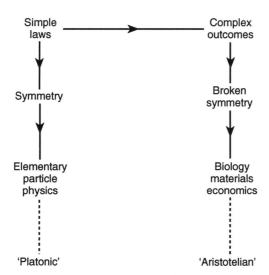

Fig. 4.2. The interrelationship between the laws of Nature and the outcomes of those laws, together with the scientific disciplines that focus primarily upon them. The left-hand thread from the laws of Nature is the Platonic tradition, whilst the right-hand line focuses upon temporal development and complicated outcomes and is associated with the Aristotelian perspective upon the world.

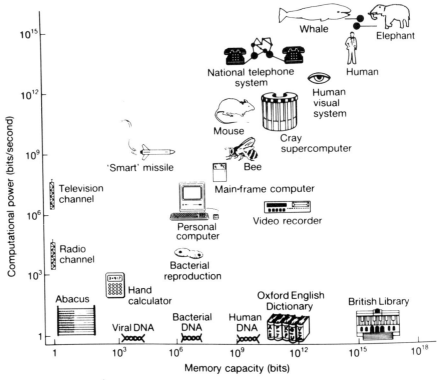

Fig. 4.3. A display, adapted from Hans Moravec,[12] showing complex organized structures in terms of their volume (in terms of capacity for storing information) and processing power (the speed at which they can change that information). We recognize the most complex structures are those in the top right-hand corner of the graph.

We see that the structure of the world around us cannot be explained by the laws of Nature alone. The broken symmetries around us may not allow us to deduce the underlying laws, and a knowledge of those laws may not allow us to deduce the permitted outcomes. Indeed, the latter state of affairs is not uncommon in fundamental physics, and is displayed by the current state of superstring theories. Theoretical physicists believe they have the laws (that is, the mathematical equations) but they are unable to deduce the outcomes of those laws (that is, find the solutions to the equations). Thus we see that whilst Theories of Everything may be necessary to understand the world we see around us, they are far from sufficient.

Of those complex outcomes of the laws of Nature, by far the most interesting are those that display what has become known as organized complexity. A selection of these are displayed in Fig. 4.3 in terms of their size (gauged by information storage capacity) versus their ability to process information (the speed with which they can change one list of numbers into another list).

Increasingly complex entities arise as we proceed up the diagonal, where increasing information storage capability grows hand-in-hand with the ability to transform that information into new forms. These complex systems are typified by the presence of feedback, self-organization, and purposeful behaviour. There may be no limit to the complexity of the entities that can exist further and further up the diagonal. Thus, for example, a complex phenomenon like high-temperature superconductivity,[13] which relies upon a very particular mixture of materials being brought together under special conditions, might never have been manifested in the Universe before the right mixtures were made on Earth in 1987. It is most unlikely that these mixtures occur naturally in the Universe, and so that variety of complexity called 'high-temperature superconductivity' relies upon that other variety of complexity called 'intelligence' to act as a catalyst for its creation. Moreover, we might speculate that there exist new types of 'law' or 'principle' that govern the existence and evolution of complexity defined in some abstract sense.[14] These rules might be quite different from the laws of the particle physicist and they might not be based upon symmetry and invariance, but upon principles of logic and information processing.

The defining characteristic of the structures in Fig. 4.3 is that they are more than the sum of their parts. They are what they are, they display the behaviour that they do, not because they are made of atoms or molecules, but because of the way in which their constituents are organized.[15] It is the circuit diagram or the neural network that is responsible for the complexity of their behaviour. The laws of electromagnetism alone are insufficient to explain the workings of a brain. We need to know how it is wired up. No Theory of Everything that the particle physicists can supply us with will shed any light upon the workings of the human brain or the nervous system of an elephant.

So far we have discussed the outcomes of the laws of Nature using rather straightforward examples, but we shall find that there are some aspects of the Universe that were once treated as unchanging parts of its constitution that have gradually begun to appear more and more like asymmetrical outcomes; indeed, the entire Universe may be one such asymmetrical outcome of an underlying law whose ultimate symmetry is hidden from us.

COSMOLOGY

One of

The greatest discovery of twentieth-century science is that the Universe is expanding; that is, the distant clusters of galaxies are receding from each other

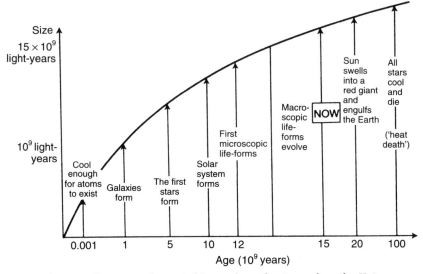

Fig. 4.4. The overall pattern of cosmic history from the time when the Universe was about a million years old until the far future. The expansion ensures that the ambient temperature of the Universe is steadily falling. Only after a particular period of time are conditions cool enough for the formation of atoms, molecules, and living complexity. In the future we expect an epoch when all the stars will have cooled and died. Thus there exists a niche of cosmic history before which life cannot evolve and after which conditions do not exist for its spontaneous evolution and persistence.

at a velocity that increases linearly with their distance apart.[16] As a result the Universe has a history. The average cosmic environment is continuously cooling and rarifying as the Universe expands and ages. If we reverse the sense of expansion in our mind's eye and trace the Universe backwards in time we encounter epochs at which the Universe was too hot for stars, molecules, or even atoms and nuclei to exist. Today, the Universe appears to have been expanding for about fifteen billion years (1 billion $= 10^9$). In Fig. 4.4 the general evolution of the Universe is mapped out.

Biologists believe that *spontaneously* evolved life must be carbon-based.[17] Given the prior existence of carbon-based life then all manner of other life forms are undoubtedly possible. We appear to be aiding the evolution of 'artificial' silicon-based life-forms at present, although they exploit the subtleties of silicon physics rather than silicon chemistry about which science fiction writers used to speculate. Carbon, and all the other biologically interesting elements heavier than helium, originate from nuclear reactions in the interiors of stars after about ten billion years of nuclear evolution. Our knowledge of the evolution of stars also teaches us that stable stars like the Sun will eventually exhaust their reserves of nuclear fuel and undergo a dramatic explosive phase of evolution before finally 'dying', unable to generate heat and energy. Hence there exists a narrow niche of cosmic history in which the spontaneous evolution of life must occur if it is ever to occur. At early times, the building blocks necessary for complexity do not exist; at later times, there are no stable hydrogen-burning stars to provide a steady environment. Obviously, we find ouselves living at a time in the history of the Universe that lies within this niche of life. This is a necessary consequence of our own existence.

The lesson that we should learn from this discussion is that the large-scale structure of the Universe is connected to the existence of living observers within it.[17] We can dramatize this by posing the question: why is the Universe so big? The visual horizon of the Universe is fifteen billion light years away. Within the volume that it encompasses there are about one hundred billion galaxies like the Milky Way, each containing about a hundred billion stars like the sun, and much else besides. Do we really need so much Universe? Could we not make do with a smaller economy-sized version?

We have seen that, in order for the building blocks of any form of organized complexity to be present in the Universe, we require at least ten billion years for the stars to produce and disperse elements heavier than helium throughout the Universe. So, because the Universe is expanding, it must be at least ten billion light years in size in order that it can contain 'observers' like ourselves. Thus the large size of the Universe should not surprise us: it is a necessary prerequisite for our own existence. Were the Universe to be just the size of the Milky Way galaxy, containing one hundred billion suns, then it would have been expanding for little more than a month—barely enough time to settle your credit card bill, let alone evolve complex observers.

This discussion highlights an important realization by cosmologists. They have recognized that 'observers' can exist only at particular times during the evolution of the Universe and, indeed, only in universes that are old enough and big enough to produce biologically useful elements (that is, those heavier than helium). One can take this consideration a little further and add that observers can exist only in particular places in the Universe where conditions are cool enough and stable enough for their evolution and persistence. These considerations are of importance if there exists any random factor in the initial make-up or early evolution of the Universe which creates conditions that differ significantly from place to place. If so, then we need to take into account that our view of the Universe is biased by the fact that we necessarily observe it from one of those regions which satisfy the necessary conditions for the evolution of complexity.

One example of this state of affairs is provided by the most general edition of the inflationary universe theory.[18] This is an elaboration of the standard big-bang picture of the expanding universe, which proposes that during the very early stages of the expansion the expansion *accelerated* for a brief period. The standard (non-inflationary) big-bang model manifests expansion that decelerates at all times, regardless of whether the expansion continues forever or collapses to a big crunch in the future. This brief period of inflation has a variety of interesting consequences: it can guarantee that the visible universe possesses many of the mysterious properties that are observed. The accelerated phase would arise if certain types of matter existed during the very hot early stages of the Universe. If this matter (which we call the 'inflation field') moved slowly enough, then it would drive a period of inflationary expansion. At the very early time when this is expected to occur, light would have had time to travel only about 10^{-25} cm since the expansion began. Thus, within each spherical region of this diameter, there should exist a smooth correlated environment; but conditions could be quite different over more widely separated dimensions, because there would not have been enough time for signals to travel between different regions. The situation is represented in Fig. 4.5.

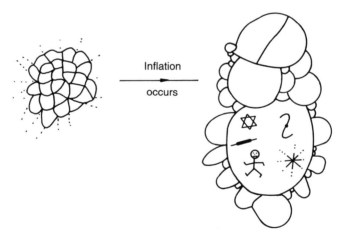

Fig. 4.5. Small causally connected domains, existing near the beginning of the expanding Universe, inflate separately by different amounts. We must find ourselves in one of the domains that expands to more than about ten billion light years in size. Only these large, old domains have enough space and time for the evolution of stars and the ensuing production of elements heavier than helium that are necessary for the evolution of complex structures.

In each of those causally disjoint regions, the inflation field would evolve differently—reflecting the differing conditions and starting states within different regions. As a result, separate regions would accelerate, or 'inflate', by different amounts in some random fashion. Some regions would inflate by large amounts and remain expanding for more than ten billion years. Only in these regions would the formation and evolution of stars, heavy elements, and organized complexity be possible. We necessarily live in one of those regions.

THE VISIBLE UNIVERSE

We should draw a distinction between 'the Universe', which might be infinite in extent, and the 'visible universe'—that part of the Universe from which we have had time to receive light signals since the beginning of the expansion about fifteen billion years ago. The visible universe is finite and steadily getting bigger as the Universe ages. All our observations of the Universe are confined to the subset of the whole that is the visible part. Now, take the region constituting the visible universe today, reverse the expansion of the Universe, and determine the size of that region which, at any given time in the past, is going to expand to become our visible universe today.[1] When the Universe was about 10^{-35} s old the present visible universe was contained with a sphere less than one centimetre in radius, as in Fig. 4.6.

This has a number of important consequences for any quest for a Theory of Everything. In particular, as can be seen from Fig. 4.7, our visible universe is the expanded image of an infinitesimal part of the initial conditions.

We can never know what the entire initial data space looked like. Moreover, whilst it is fashionable among cosmologists to attempt to formulate grand 'principles' that specify the initial conditions (for example, Roger Penrose's 'minimum Weyl entropy condition',[19] Alex Vilenkin's 'outgoing wave condition',[20] or James Hartle and Stephen Hawking's 'no boundary condition'[21]), these principles specify averaged conditions over the entire initial data space. Ultimately, these conditions will be quantum-statistical in character. However, such principles could never be tested—even with all the information available in the visible universe. The entire visible universe is determined by a tiny part of the initial data space, which may well be atypical in certain respects, in order that it should satisfy the conditions necessary for the subsequent evolution of observers. Global principles about the initial state of the Universe, even if correct, may be of little use in understanding the structure of the visible part of the Universe because it evolves from an idiosyncratic part of an infinite span of initial conditions.

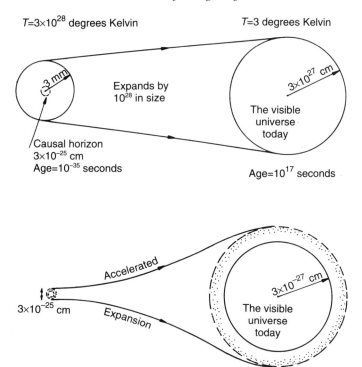

Fig. 4.6. The past history of the region defining our visible universe today. When the Universe was 10^{-35} s old, this was 1 cm in size. By contrast, the visible universe at that time was 10^{-25} cm in radius. Thus, our visible universe evolved from a vast number of causally disjoint regions. Why then does it look so uniform from place to place over large scales? Why does the background radiation field have the same temperature in every direction to within a few parts in one hundred thousand? The inflationary universe picture can resolve this dilemma by accelerating the early expansion of the Universe for a brief period. This more rapid expansion enables the visible universe today to have expanded from a region of radius only 10^{-25} cm, or smaller, at a time of 10^{-35} s. This particular time is used for illustrative purposes. Inflation does not have to occur at this particular time, although it could well have done so.

CONSTANTS OF NATURE

The constants of physics—quantities like the mass of an electron or its electric charge—have traditionally been regarded as unchanging attributes of the Universe. An acid test of any Theory of Everything would be its ability to predict the values of these constants.[22]

The values of the constants of Nature are what endow the Universe with its coarse-grained structure. The sizes of stars and planets, and to some extent

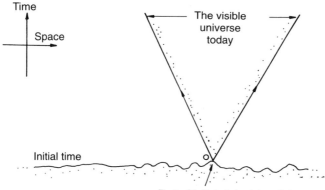

Fig. 4.7. A space and time diagram showing the development of our visible universe from its initial conditions. The structure of our entire visible universe is determined by conditions at the point o. These conditions may need to be atypical in order to permit the subsequent evolution of observers and their detailed local form will not be determined by any global principle which determines the initial conditions quantum-statistically.

also that of people, are determined by the values of these constants.[23] In addition, it has gradually been appreciated that the necessary conditions for the evolution of organized complexity in the Universe are dependent upon a large number of remarkable 'coincidences' between the values of different constants. Were the constants governing electromagnetic or nuclear interactions to be very slightly altered in value, then the chain of nuclear reactions that produce carbon in the Universe would fail to do so; change the constants a little more and neither atoms, nor nuclei, nor stable stars, could exist.[17,23–26] Hence, we believe that only those universes in which the constants of Nature lie in a very narrow range (including, of course, the actual values) can give rise to observers of any sort.

In recent years, our attempts to develop a quantum cosmological model have uncovered the remarkable possibility that the constants of Nature may be predictable but their values may only be determined quantum-statistically by the space–time fabric of the Universe.[27]

If we relinquish the idea that the topological structure of the Universe is a smooth ball, then we picture it as a crenellated structure with a network of wormholes connecting it to itself and to other extended regions of space–time, as pictured in Fig. 4.8. The size of the throats of the wormholes is of order 10^{-33} cm. Now, this turns out to be something more interesting than simply generalization for its own sake. It appears that the values of the observed constants of Nature on each of the large regions of space–time may be determined quantum-statistically by the network of wormhole connections. Thus, even if those constants had their values determined initially by the logical strait-jacket of some unique Theory of Everything, these values would be shifted so that today they would be observed to have values given by a calculable quantum-probability distribution. There is, however, a subtlety here.[1] We might think that it is the business of science to compare the most probable value predicted with those observed. But why should the observed universe display the most probable value, in any sense of the world 'probable'? We know that observers can evolve only if the constants lie in a very narrow range; so the situation will be as shown in Fig. 4.9.

We are interested, not in the most probable value of a constant, but only in the conditional probability of the constant taking a value that subsequently permits living observers to evolve. This may be very different from the unconditioned 'most probable' value.

This type of argument can be extended to consider the dimensionality of the part of space that becomes large. Superstring theories naturally predict that the Universe possesses more than three dimensions of space,[8,9] but it is expected that only some of these dimensions will expand to become very large. The rest must remain confined and imperceptibly small. Now we might ask why it should be *three* dimensions that grow large. Is this the only logical possibility or just a random symmetry-breaking that could have fallen out in another way—or perhaps *has* fallen out in different ways all over the Universe,

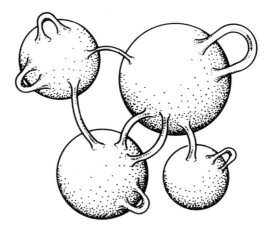

Fig. 4.8. Wormholes connecting different large sub-regions of the Universe to themselves by wormhole 'handles' and to other such regions by wormhole 'throats'.

so that regions beyond our horizon may have four or more large dimensions? Again, we know that the dimensionality of space is critically linked to the likelihood of observers evolving within it.[28] (For instance, there can be no atoms or stable gravitationally bound orbits in more than three dimensions.) Considerations similar to those we have just introduced with regard to the probabilistic prediction of the constants will also apply to any random symmetry-breaking process that determines the number of space dimensions that become large.

These examples reiterate that the output of a Theory of Everything is not immediately comparable with the observed universe because, if there exists a random element in the initial structure of the Universe (and quantum theory makes this inevitable), we may observe aspects of things that are not typical. They are conditioned by the necessary conditions for the evolution of complexity. The observed universe should be thought of as being an outcome of the laws of Nature, because we observe only part of what might be an infinite whole. Whilst a Theory of Everything is necessary to understand the observed part of the universe, it is far from sufficient. We need to understand the role and nature of the initial conditions; to understand how our visible universe may be atypical in order to satisfy the necessary conditions for the evolution of observers; and to understand how complex symmetries break to allow the subsequent evolution of organized complexity in the Universe.

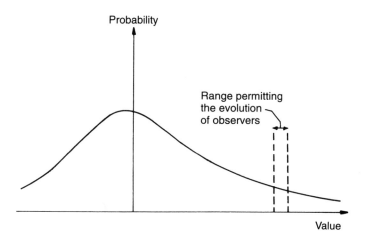

Fig. 4.9. A hypothetical prediction for the measured probability of a constant of Nature when the Universe is large and old (i.e. now). The small range of values for which living complexity can exist is also shown. The latter range may be far from the most probable value predicted by the wormhole fluctuations. We should be interested only in the most probable values which permit observers subsequently to evolve.

WHAT ARE THE ULTIMATE RULES OF THE GAME?

Physicists tend to believe that a Theory of Everything will be some set of equations governing entities like points or strings and will be equivalent to the preservation of some underlying symmetry. This is an extrapolation in the direction in which particle physics has been moving for some time. A key assumption of such a picture is that it regards the laws of physics as being the bottom line and assume that these laws govern a world of point particles or strings (or other exotica) that is a continuum. Another possibility is that the Universe is not at root a great symmetry but a computation. The ultimate laws of Nature may be akin to software running upon the hardware provided by elementary particles and energy. The laws of physics might then be derived from some more basic principles governing computation and logic. This view might have radical consequences for our appreciation of the subtlety of Nature, for it seems to require that the world is at root discontinuous, like a computation, rather than a continuum. This makes the Universe a much more complicated place. If we count the number of discontinuous changes that can exist, we find that there are infinitely many more of them than there are continuous changes. By regarding the bedrock structure of the Universe as a continuum we may not just be making an approximation but an infinite simplification.

REFERENCES

1. Barrow, J. D. (1991). *Theories of Everything: the quest for ultimate explanation.* Oxford University Press (Vintage (paperback) 1992).
2. Long, C. H. (1963). *Alpha: the myths of creation.* G. Braziller, New York.
3. Chaitin, G. (1987). *Algorithmic information theory.* Cambridge University Press.
4. Chaitin, G. (1980). Randomness in arithmetic. *Scientific American*, July, 80–5.
5. Barrow, J. D. (1992). *Pi in the sky: counting, thinking, and being.* Oxford University Press.
6. Barrow, J. D. (1988). *The world within the world.* Oxford; Feynman, R. (1965). *The character of physical law.* MIT Press, Boston.
7. Pagels, H. (1985). *Perfect symmetry.* Michael Joseph, London; Zee, A. (1986). *Fearful symmetry: the search for beauty in modern physics.* Macmillan; Weinberg, S. (1983). *The discovery of sub-atomic particles.* W. H. Freeman.
8. Green, M., Schwarz, J., and Witten, E. (1987). *Superstring theory* (2 vols). Cambridge University Press.
9. Green, M. (1986). Superstrings. *Scientific American*, September, p. 48; Bailin, D. (1989). Why superstrings?. *Contemporary Physics*, **30**, 237; Davies, P. C. W. and Brown, J. R. (eds.) (1988). *Superstrings: a Theory of Everything?.* Cambridge University Press.

10. Barrow, J. D. (1991). Platonic relationships in the Universe. *New Scientist*, 20 April, p. 40.

11. Gleick, J. (1987). *Chaos: making a new science*. Viking, New York; Stewart, I. (1989). *Does God play dice: the mathematics of chaos*. Blackwell, Oxford; Mandelbrot, B. (1982). *The fractal geometry of Nature*. W. H. Freeman, San Francisco.

12. Moravec, H. (1988). *Mind children*. Harvard University Press.

13. Gough, C. (1991). Challenges of High-T_c. *Physics World*, December, 26.

14. Lloyd, S. and Pagels, H. (1988). Complexity as thermodynamic depth. *Annals of Physics* (New York), **188**, 186–213.

15. Davies, P. C. W. (1987). *The cosmic blueprint*. Heinemann, London.

16. Weinberg, S. *The first three minutes*. André Deutsch, London; Barrow, J. D. and Silk, J. (1983). *The left hand of creation*. Basic Books, New York.

17. Barrow, J. D. and Tipler, F. J. (1986). *The anthropic cosmological principle*. Oxford University Press.

18. Guth, A. (1981). The inflationary universe: a possible solution to the horizon and flatness problems. *Physical Review* D **23**, 347–56; Guth, A. and Steinhardt, P (1984). The inflationary universe. *Scientific American*, May 1984, pp. 116–20; Barrow, J. D. (1988). The inflationary universe: modern developments. *Quarterly Journal of the Royal Astronomical Society*, **29**, 101–17.

19. Penrose, R. (1989). *The Emperor's new mind*. Oxford University Press.

20. Vilenkin, A. (1982). Boundary conditions in quantum cosmology. *Physical Review*, D **33**, 3560–9.

21. Hartle, J. B. and Hawking, S. W. (1983). The wave function of the Universe, *Physical Review* D **28**, 2960–75; Hawking, S. W. (1988). *A brief history of time*. Bantam Books, London.

22. An infamous, completely unsuccessful, programme of this sort was Arthur Eddington's quest for a 'Fundamental Theory'; see Eddington, A. S. (1946). *Fundamental theory*. Cambridge University Press, for an edited posthumous presentation, and Eddington, A. S. (1935). *New pathways in science*. Cambridge University Press, for a more popular account of its early progress. See also ref. 17 for analysis.

23. Barrow, J. D. (1990). The mysterious law of large numbers. In *Modern cosmology in retrospect* (ed. S. Bergia and B. Bertotti), pp. 67–93 Cambridge University Press.

24. Carr, B. J. and Rees, M. J. (1969). The anthropic principle and the structure of the physical world. *Nature*, **278**, 605–8.

25. Davies, P. C. W. (1982). *The Accidental Universe*. Cambridge.

26. Gribbin, J. and Rees, M. J. (1989). *Cosmic coincidences: dark matter, mankind, and anthropic cosmology*. Bantam Books, London.

27. Coleman, S. (1988). Why there is something rather than nothing. *Nuclear Physics*, B **310**, 643–68; Hawking, S. W. (1988). Wormholes in space–time. *Physical Review*, D **37**, 904–10; Hawking, S. W. (1990). Baby universes. *Modern Physics Letters*, A **5**, 145–55.

28. Barrow, J. D. (1983). Dimensionality. *Philosophical Transactions of the Royal Society of London*, A **310**, 337–46.

CHAPTER FIVE

Intertheoretic reduction: a neuroscientist's field guide

PAUL M. CHURCHLAND AND PATRICIA S. CHURCHLAND

'Reductionism' is a term of contention in academic circles. For some it connotes a right-headed approach to any genuinely scientific field, an approach that seeks intertheoretic unity and real systematicity in the phenomena. It is an approach to be vigorously pursued and defended. For others, it connotes a wrong-headed approach that is narrow-minded and blind to the richness of the phenomena. It is a bullish instance of 'nothing-but-ery', insensitive to emergent complexity and higher-level organization. It is an approach to be resisted.

The latter reaction is most often found within the various social sciences, such as anthropology, sociology, and psychology. The former attitude is most often found within the physical sciences, such as physics, chemistry, and molecular biology. Predictably then, the issue of reductionism is especially turbulent at the point where these two intellectual rivers meet: in the discipline of modern neuroscience.

The question at issue is whether it is reasonable to expect, and to work toward, a reduction of all psychological phenomena to neurobiological and neurocomputational phenomena. A large and still respectable contingent within the academic community remains inclined to say 'no'. Their resistance is principled. Some point to the existence of what philosophers call *qualia*—the various subjective qualitative characters displayed in our sensations: think of pain, the smell of a rose, the sensation of redness, and so forth. These qualia, it is held, are beyond the possibility of any materialist explanation or reduction.[1,2] Others point to the semantic content or *intentionality* of our thoughts, and make a similar claim about its irreducibility.[3-5] Others claim that the most important aspects of human behaviour are explicable only in terms of high-level emergent properties and their correlative regularities, properties that irreducibly encompass the social level, properties such as loyalty to a moral ideal, perception of a political fact, or the recognition of a personal betrayal.[6,7] Yet others see a conflict with the important and deeply entrenched idea of human freedom.[3] Finally, some materialists raise what is called the problem of *multiple instantiation*. They point to the presumed fact that conscious intelligence could be sustained by physical systems other than the biochemistry peculiar to humans—by a system of transistors, for example—just as a nation's financial economy can be sustained by tokens other than silver coins and paper bills. But no one thinks that macro-economics can be reduced to the chemistry of metals and paper. So why think that psychology should be reducible to the neurobiology of terrestrial humans?[8]

Our aim here is threefold. First, we shall try to provide a useful overview of the general nature of intertheoretic reduction, as it appears in the many examples to be found in the history of science. Expanding our horizons here is important, since little is to be learned from simply staring long and hard at the problematic case at issue, namely, the potential reduction of psychological phenomena to neural phenomena. Instead, we need to look at cases where the dust has already settled and where the issues are already clear. Second, we shall identify the very real virtues that such cases display, and the correlative vices to be avoided. And finally, we shall attempt to apply these historical lessons to the case here at issue—cognitive neuroscience—and we shall try to meet the salient objections listed above.

INTERTHEORETIC REDUCTION: SOME PROTOTYPICAL CASES

Since nothing instructs like examples, let us briefly examine some. One of the earliest cases of intertheoretic reduction on a grand scale was the reduction of Kepler's three laws of astronomical motion by the newly minted mechanics of Isaac Newton. Kepler's theory was specific to the motions of the solar planets, but Newton's theory at least purported to be the correct account of bodily motions in general. It was therefore a great triumph when Newton showed that one could deduce all three of Kepler's laws from his own theory, given only the background assumption that the mass of any planet is tiny compared to the great mass of the Sun.

Kepler's account thus turned out to be just a special case or a special application of Newton's more encompassing account. And astronomical motions turned out to be just a special instance of the inertial and force-

---

Kepler's three planetary laws are:

(1) All planets move in elliptical orbits with the Sun at one focus;
(2) A given planet always sweeps out equal areas in equal times;
(3) The square of planet's period is proportional to the cube of its mean orbital radius.

Newton's three laws of motion are:

(1) Inertial motion is constant and rectilinear; (2) Acceleration = force/mass; (3) For any change in momentum something suffers an equal and opposite change in momentum. To these laws we must add his gravitation law:

$$F = Gm_1m_2/R^2$$

---

governed motions of massive bodies in general. The divine or supernatural character of the heavens was thereby lost for ever. The sublunary and the superlunary realms were thereby united as a single domain in which the same kinds of objects were governed by one and the same set of laws.

Newton's mechanics also provides a second great example of intertheoretic reduction, one that did not emerge until the nineteenth century. If his mechanics successfully comprehends motion at both the astronomical and the human-sized scales, then what, it was asked, about motions at the microscopic scale? Might these be accounted for in the same way?

The attempts to construct such an account produced another unification, one with an unexpected bonus concerning the theory of heat. If we assume that any confined body of gas consists in a swarm of submicroscopic corpuscles bounding around inside the container according to Newton's three laws, then we can deduce a law describing the pressure they will collectively exert on the container's walls by repeatedly bouncing off them. This 'kinetic' law has the form

$$PV = 2n/3 \cdot mv^2/2$$

This law had the same form as the then already familiar 'ideal gas law',

$$PV = \mu R \cdot T$$

(Here P is pressure and V is volume.) Although they are notationally different, the expressions $2n/3$ and μR both denote the amount of gas present in the container (n denotes the number of molecules in the container; μ denotes the fraction of a mole). The only remaining difference, then, is that the former law has an expression for the kinetic energy of an average corpuscle ($mv^2/2$) in the place where the latter has an expression for temperature (T). Might the phenomenon we call 'temperature' thus *be* mean kinetic energy at the molecular level? This striking convergence of principle, and many others like it, invited Bernoulli, Joule, Kelvin, and Boltzmann to say 'yes'. As matters were further pursued, mean molecular kinetic energy turned out to have *all* the causal properties that the classical theory had been ascribing to temperature. In short, temperature turned out to *be* mean molecular kinetic energy. Newtonian mechanics had another reductive triumph in hand. Motion at all three scales was subsumed under the same theory, and a familiar phenomenal property, temperature, was reconceived in a new and unexpected way.

It is worth emphasizing that this reduction involved identifying a familiar phenomenal property of common objects with a highly unfamiliar microphysical property. (By 'phenomenal' we mean a property one can reliably discriminate in experience, but where one is unable to articulate, by reference to yet simpler discriminable elements, just how one discriminates that property.) Evidently, reduction is not limited to conceptual frameworks hidden away in the theoretical stratosphere. Sometimes the conceptual framework that becomes subsumed by a deeper vision turns out to be a familiar piece of our common-sense framework, a piece whose concepts are regularly applied in casual observation on the basis of our native sensory systems. Other examples are close at hand: before Newton, sound had already been identified with compressional waves in the atmosphere, and pitch with wavelength, as part of the larger reduction of common-sense sound and musical theory to mechanical acoustics. A century and a half after Newton, light and its various colours were identified with electromagnetic waves and their various wavelengths, within the larger reduction of geometrical optics by electromagnetic theory, as outlined by Maxwell in 1864. Radiant heat, another common-sense observable, was similarly reconceived as long-wavelength electromagnetic waves in a later articulation of the same theory. Evidently, the fact that a property or state is at the prime focus of one of our native discriminatory faculties does not mean that it is exempt from possible reconception within the conceptual framework of some deeper explanatory theory.

This fact will loom larger later in this contribution. Let us first explore some further examples of intertheoretic reduction. The twentieth-century reduction of classical (valence) chemistry by atomic and subatomic (quantum) physics is another impressive case of conceptual unification. Here the structure of an atom's successive electron shells, and the character of stable regimes of electron-sharing between atoms, allowed us to reconstruct, in a systematic and thus illuminating way, the electronic structure of the many atomic elements, the classical laws of valence-bonding, and the gross structure of the periodic table. As often happens in intertheoretic reductions, the newer theory also allowed us to explain much that the old theory had been unable to explain, such as the specific heat capacities of various substances and the interactions of chemical compounds with light.

This reduction of chemistry to physics is notable for the further reason that it is not yet complete, and probably never will be. For one thing, given the combinatorial possibilities here, the variety of chemical compounds is effectively endless, as are their idiosyncratic chemical, mechanical, optical, and thermal properties. And, for another, the calculation of these diverse properties from basic quantum principles is computationally daunting, even when we restrict ourselves to merely approximate results, which for mathematical reasons alone we generally must. Accordingly, it is not true that all chemical knowledge has been successfully reconstructed in quantum-mechanical terms. Only the basics have, and then only in approximation. But our experience here firmly suggests that quantum physics has indeed managed to grasp the underlying elements of chemical reality. We thus expect that any particular part of chemistry can be approximately reconstructed in quantum-mechanical terms, when and if the specific need arises.

The preceding examples make it evident that intertheoretic reduction is at bottom a relation between two distinct *conceptual frameworks* for describing the phenomena, rather than a relation between two distinct domains of phenomena. The whole point of a reduction, after all, is to show that what we thought to be two domains is actually one domain, though it may have been described in two (or more) different vocabularies.

Perhaps the most famous reduction of all is Einstein's twentieth-century reduction of Newton's three laws of motion by the quite different mechanics of the Special Theory of Relativity (STR). STR subsumed Newton's laws in the following sense. If we make the (false) assumption that all bodies move with velocities much less than the velocity of light, then STR entails a set of laws for the motion of such bodies, a set that is experimentally indistinguishable from Newton's old set. It is thus no mystery that those old Newtonian laws seemed to be true, given the relatively parochial human experience they were asked to account for.

But while those special-case STR laws may be experimentally indistinguishable from Newton's laws, they are logically and semantically quite different from Newton's laws: they ascribe a significantly different family of features to the world. Specifically, in every situation where Newton ascribed an intrinsic property to a body (e.g. mass, or length, or momentum, and so forth), STR ascribes a *relation*, a *two*-place property (e.g. *x* has a mass-relative-to-an-inertial-frame-F, and so on), because its portrait of the universe and what it contains (a unitary four-dimensional space–time continuum with four-dimensional world-lines) is profoundly different from Newton's.

Here we have an example where the special-case resources and deductive consequences of the new and more general theory are not identical, but merely *similar*, to the old and more narrow theory it purports to reduce. That is to say, the special-case reconstruction achieved within the new theory parallels the old theory with sufficient systematicity to explain why the old theory worked as well as it did in a certain domain, and to demonstrate that the old theory could be displaced by the new without predictive or explanatory loss within the old theory's domain; and yet the new reconstruction is not perfectly isomorphic to the old theory. The old theory turns out not just to be narrow, but to be false in certain important respects. Space and time are not distinct, as Newton assumed, and there simply are no intrinsic properties such as mass and length that are invariant over all inertial frames.

The trend of this example leads us toward cases where the new and more general theory does not sustain the portrait of reality painted by the old theory at all, even as a limiting special case or even in its roughest outlines. An example would be the outright displacement, with reduction, of the old phlogiston theory of combustion by Lavoisier's oxygen theory of combustion. The older theory held that the combustion of any body involved the loss of a spirit-like substance, phlogiston, whose pre-combustion function it was to provide a noble wood-like or metal-like character to the baser ash or calx that is left behind after the process of combustion is complete. It was the 'ghost' that gave metal its form. With the acceptance of Lavoisier's contrary claim that a sheerly material substance, oxygen, was somehow being absorbed during combustion, phlogiston was simply eliminated from our overall account of the world.

Other examples of theoretical entities that have been eliminated from serious science include caloric fluid, the rotating crystal spheres of Ptolemaic astronomy, the four humours of medieval medicine, the vital spirit of pre-modern biology, and the luminiferous aether of pre-Einsteinian mechanics. In all these cases, the newer theory did not have the resources necessary to reconstruct the furniture of the older theory or the laws that supposedly governed its behaviour; but the newer theory was so clearly superior to the old as to displace it regardless.

At one end of the spectrum then, we have pairs of theories where the old is smoothly reduced by the new, and the ontology of the old theory (that is, the set of things and properties that it postulates) survives, although redescribed, perhaps, with a new and more penetrating vocabulary. Here we typically find claims of cross-theoretic identity, such as 'Heat is identical with mean molecular kinetic energy' and 'Light is identical with electromagnetic waves'. In the middle of the spectrum, we find pairs of theories where the old ontology is only poorly mirrored within the vision of the new, and it 'survives' only in a significantly modified form. Finally, at the other end of the spectrum we find pairs where the older theory, and its old ontology with it, is eliminated entirely in favour of the more useful ontology and the more successful laws of the new.

Before closing this brief survey, it is instructive to note some cases where the older theory is neither subsumed under nor eliminated by the aspirant and allegedly more general theory. Rather, it successfully resists the takeover attempt, and proves not to be just a special case of the general theory at issue. A clear example is Maxwell's electromagnetic theory (hereafter, EM theory). From 1864 to 1905, it was widely expected that EM theory would surely find a definitive reduction in terms of the mechanical properties of an all-pervading aether, the elastic medium in which EM waves were supposedly propagated. Though never satisfactorily completed, some significant attempts at reconstructing EM phenomena in mechanical terms had already been launched. Unexpectedly, the existence of such an absolute medium of luminous propagation turned out to be flatly inconsistent with the character of space and time as described in Einstein's 1905 Special Theory of Relativity. EM theory thus emerged as a fundamental theory in its own right, and not just as a special case of mechanics. The attempt at subsumption was a failure.

A second example concerns the theory of stellar behaviour accumulated by classical astronomy in the late nineteenth century. It was widely believed that the pattern of radiative behaviour displayed by a star would be adequately explained in mechanical or in chemical terms. It became increasingly plain, however, that the possible sources of chemical and mechanical energy available to any star would sustain their enormous outpourings of thermal and luminous energy for only a few tens of millions of years. This limited time-scale was at odds with the emerging geological evidence of a history numbered in *billions* of years. Geology notwithstanding, Lord Kelvin himself was prepared to bite the bullet and declare the stars to be no more than a few tens of millions of years old. The conflict was finally resolved when the enormous energies in the atomic nucleus were discovered. Stellar astronomy was eventually reduced, and very beautifully, but by quantum physics rather than by mere chemistry of mechanics. Another reductive attempt had failed, though it was followed by one that succeeded.

THE LESSONS FOR NEUROSCIENCE

Having seen these examples and the spectrum of cases they define, what lessons should a neuroscientist draw? One lesson is that intertheoretic reduction is a normal and fairly commonplace event in the history of science. Another lesson is that genuine reduction, when you can get it, is clearly a good thing. It is a good thing for many reasons, reasons made more powerful by their conjunction.

First, by being displayed as a special case of the (presumably true) new theory, the old theory is thereby vindicated, at least in its general outlines, or at least in some suitably restricted domain. Second, the old theory is typically corrected in some of its important details, since the reconstructed image is seldom a perfect mirror image of the old theory, and the differences reflect improvements in our knowledge. Third, the reduction provides us with a much deeper insight into, and thus a more effective control over, the phenomena within the old theory's domain. Fourth, the reduction provides us with a simpler overall account of nature, since apparently diverse phenomena are brought under a single explanatory umbrella. And fifth, the new and more general theory immediately inherits all the evidence that had accumulated in favour of the older theory it reduces, because it explains all the same data.

It is of course a bad thing to try to force a well-functioning old theory into a procrustean bed, to try to effect a reduction where the aspirant reducing theory lacks the resources to do reconstructive justice to the target old theory. But whether or not the resources are adequate is seldom clear beforehand, despite people's intuitive convictions. And even if a reduction is impossible, this may reflect the old theory's radical falsity instead of its fundamental accuracy. The new theory may simply eliminate the old, rather than smoothly reduce it. Perhaps folk notions such as 'beliefs' and 'the will', for example, will be eliminated in favour of some quite different story of information storage and behaviour initiation.

The fact is that in the case of neuroscience/psychology there are conflicting indications. On the one side, we should note that the presumption in favour of an eventual reduction (or elimination) is far stronger than it was in the historical cases just examined. For, unlike the earlier examples of light, or heat, or heavenly motions, in general terms we already know how psychological phenomena arise: they arise from the evolutionary and ontogenetic articulation of matter; more specifically, from the articulation of biological organization. We therefore *expect* to understand the former in terms of the latter. The former is produced by the relevant articulation of the latter.

But there are counter-indications as well, and this returns us at last to the five objections with which we opened. From the historical perspective outlined above, can we say anything useful about those objections to reduction? Let us take them in sequence.

The first concerns the possibility of explaining the character of our subjective sensory qualia. The negative arguments here all exploit the very same theme, namely our inability to imagine how any possible story about the objective nuts and bolts of neurones could ever explain the inarticulable subjective phenomena at issue. Plainly this objection places a great deal of weight on what we can and cannot imagine, as a measure of what is and is not possible. It places more, clearly, than the test should bear. For who would have imagined, before James Clerk Maxwell, that the theory of charged pith balls and wobbling compass needles could prove adequate to explain all the phenomena of light? Who would have thought, before Descartes, Bernoulli, and Joule, that the mechanics of billiard balls would prove adequate to explain the *prima facie* very different phenomenon of heat? Who would have found it remotely plausible that the pitch of a sound is a frequency, in advance of a general appreciation that sound itself consists of a train of compression waves in the atmosphere?

We must remember that a successful intertheoretic reduction is typically a complex affair, since it involves the systematic reconstruction of all or most of the old conception within the resources of the new conception. And not only is it complex; the reconstruction is often highly surprising. It is not something that we can reasonably expect anyone's imagination to think up or comprehend on rhetorical demand, as in the question, 'How could *As possibly* be nothing but *Bs*?'

Besides, this rhetorical question need not stump us if our imagination is informed by recent theories of sensory coding. The idea that taste sensations are coded as a four-dimensional vector of spiking frequencies (corresponding to the four types of receptor on the tongue) yields a representation of the space of humanly possible tastes which unites the familiar tastes according to their various similarities, differences, and other relations such as between-ness.[9] Land's retinex theory of colour vision suggests a similar arrangement for our colour sensations, with similar virtues.[10] Such a theory also predicts the principal forms of colour blindness, as when an individual's three-dimensional colour space is reduced to two dimensions by the loss of one of the three classes of retinal cones.

Here we are already reconstructing some of the features of the target phenomena in terms of the new theory. We need only to carry such a reconstruction through, as in the historical precedents of the objective phenomenal properties noted earlier (heat, light, pitch). Some things may indeed be inarticulably phenomenal in character, because they are the target of one of our basic discriminatory modalities. But that in no way makes them immune to an illuminating intertheoretic reduction. History already teaches us the contrary.

The second objection concerns the meaning, or semantic content, or intentionality of our thoughts and other mental states. The anti-reductionist arguments in this area are very similar to those found in the case of qualia. They appeal to our inability to imagine how meaning could be just a matter of how signals interact or how inert symbols are processed (Searle[4,5]; for a rebuttal, see Churchland and Churchland.[11] Searle, strictly speaking, objects only to a purely computational reduction, but that is an important option for neuroscience so we shall include him with the other anti-reductionists.) Such appeals, as before, are really arguments from ignorance. They have the form, 'I can't *imagine* how a neurocomputational account of meaningful representations could possibly work, therefore, it can't possibly work.' To counter such appeals in the short term, we need only point out this failing.

To counter them in the long term requires more. It requires that we actually produce an account of how the brain represents the external world and the regularities it displays. But that is precisely what current theories of neural network function address. According to them, real-time information about the world is coded in activation vectors, and general information about the world is coded in the background configuration of the network's synaptic weights. Activation vectors are processed by the weight-configurations through which they pass, and learning consists in the adjustment of one's global weight-configuration. These accounts already provide the resources to explain a variety of things, such as the recognition of complex objects despite partial or degraded sensory inputs, the swift retrieval of relevant information from a vast content-addressable memory, the appreciation of diffuse and inarticulable similarities, and the administration of complex sensorimotor coordination.[13] We are still too ignorant to insist that hypotheses of this sort will prove adequate to explain all of the representational capacities of mind. But neither can we insist that they are doomed to prove inadequate. It is an empirical question, and the jury is still out.

The third objection complains that what constitutes a human consciousness is not just the intrinsic character of the creature itself, but also the rich matrix of relations it bears to other humans. A reductionist account of human consciousness and behaviour, in so far as it is limited to the microscopic activities in an individual's brain, cannot hope to capture more than a small part of what is explanatorily important.

The proper response to this objection is to embrace it. Human behaviour is indeed a function of the factors cited. And the character of any individual human consciousness will be profoundly shaped by the culture in which it develops. What this means is that any adequate neuro-computational account of human consciousness must take into account the manner in which a brain comes to represent, not just the gross features of the physical world, but also the character of the other cognitive creatures with which it interacts, and the details of the social, moral, and political world in which they all live. The brains of social animals, after all, learn to be interactive elements in a community of brains, much to their cognitive advantage. We need to know how they do it.

This is a major challenge, one that neuroscientists have not yet addressed with any seriousness, nor even much acknowledged. This is not surprising. Accounting for a creature's knowledge of the spatial location of a fly is difficult enough. Accounting for its knowledge of a loved one's embarrassment, a politician's character, or a bargaining opponent's hidden agenda, represents a much higher level of difficulty. And yet we already know that artificial neural networks, trained by examples, can come to recognize and respond to the most astonishingly subtle patterns and similarities in nature. If physical patterns, why not social patterns? We confront no problem in principle here. Only a major challenge.

It may indeed be unrealistic to expect an exhaustive global account of the neural and behavioural trajectory of a specific person over any period of time. The complexity of the neural systems we are dealing with may for ever preclude anything more than useful approximations to the desired ideal account. The case of chemistry and its relation to quantum physics comes to mind. There also, the mathematics of complex dynamical systems imposes limits on how easily and accurately we can reconstruct the chemical facts from the physical principles. This means that our reduction will never be truly complete, but we rightly remain confident that chemical phenomena are nothing but the macro-level reflection of the underlying quantum physical phenomena even so. As with chemical phenomena, so with psychological phenomena.

This brings us to the fourth objection, concerning the threat that a reduction would pose to human freedom. Here we shall be brief. Whether and in what sense there is any human freedom, beyond the relative autonomy that attaches to any complex dynamical system that is partially isolated from the world, is an entirely empirical question. Accordingly, rather than struggle to show that a completed neuroscience will be consistent with this, that, or the other preconceived notion of human freedom, we recommend that we let scientific investigation *teach us* in what ways and to what degrees human creatures are 'free'. No doubt this will entail modifications for some people's current conceptions of human freedom, and the complete elimination of some others. But that is preferable to making our current confusions into a standard that future theories must struggle to be consistent with.

The fifth and final objection claims an irreducibly abstract status for psychology, on grounds that a variety of quite different physical systems could realize equally well the abstract organization that constitutes a cognitive economy. How can we reduce psychological phenomena to neurobiology, if other physical substrates might serve just as well?

The premise of this objection will probably be conceded by all of us. But the conclusion against reduction does not follow. We can see this clearly by examining a case from our own scientific history. Temperature, we claimed earlier, is identical with mean molecular kinetic energy. But strictly speaking, this is true only for a gas, where the molecules are free to move in a ballistic fashion. In a solid, where the particles oscillate back and forth, their energy is constantly switching between a kinetic and a potential mode. In a high-temperature plasma there are no molecules at all to consider, since everything has been ripped into subatomic parts. Here temperature is a complex mix of various energies. And in a vacuum, where there is no mass at all, temperature consists in the wavelength distribution—the 'black-body curve'—of the EM waves passing through it.

What these examples show us is that reductions can be domain-specific: in a gas, temperature is one thing; in a solid, temperature is another thing; in a plasma, it is a third; in a vacuum, a fourth; and so on. (They all count as 'temperatures', since they interact, and they all obey the same laws of equilibrium and disequilibrium.) None of this moves us to say that classical thermodynamics is an autonomous, irreducible science, for ever safe from the ambitions of the underlying microphysical story. On the contrary, it just teaches us that there is more than one way in which energy can be manifested at the microphysical level.

Similarly, visual experience may be one thing in a mammal, a slightly different thing in an octopus, and a substantially different thing in some possible metal-and-semiconductor android. But they will all count as visual experiences because they share some set of abstract features at a higher level of description. That neurobiology should prove capable of explaining all psychological phenomena in humans is not threatened by the possibility that some other theory, say semiconductor electronics, should serve to explain psychological phenomena in robots. The two reductions would not conflict. They would complement each other.

We have elsewhere provided more comprehensive accounts of how recent work in neuroscience illuminates issues in psychology and cognitive theory.[12,13] We conclude here with two cautionary remarks. First, while we have here been very positive about the possibility of reducing psychology to neuroscience, producing such a reduction will surely be a long and difficult business. We have here been concerned only to rebut the counsel of impossibility, and to locate the reductive aspirations of neuroscience in a proper historical context.

Second, it should not be assumed that the science of psychology will somehow disappear in the process, nor that its role will be limited to that of a passive target of neural explanation. On the contrary, chemistry has not disappeared through the quantum-mechanical explication of its basics; nor has the science of biology disappeared, despite the chemical explication of its basics. Moreover, each of these higher-level sciences has helped to shape profoundly the development and articulation of its underlying science. It will surely be the same with psychology and neuroscience. At this level of complexity, intertheoretic reduction does not appear as the sudden takeover of one discipline by another; it more closely resembles a long and slowly maturing marriage.

REFERENCES

1. Jackson, F. (1982). Epiphenomenal qualia. *Philosophical Quarterly*, **32**, 127–36.
2. Nagel, T. (1974). What is it like to be a bat? *Philosophical Review*, **83**, 435–50.
3. Popper, K. and Eccles, J. (1978). *The self and its brain*. Springer, New York.
4. Searle, J. (1980). Minds, brains, and programs. *Behavioural and Brain Sciences*, **3**, 417–57.
5. Searle, J. (1990). Is the brain's mind a computer program? *Scientific American*, **262**, 26–31.
6. Taylor, C. (1970). Mind–body identity: a side issue? In *The mind/brain identity theory* (ed. C.B. Borst), pp. 231–41. Macmillan, Toronto.
7. Taylor, C. (1987). Overcoming epistemology. In *After philosophy: end or transformation?* (ed. K. Baynes, J. Bohman, and T. McCarthy), pp. 464–88. MIT Press, Cambridge, Mass.
8. Fodor, J.A. (1975). *The language of thought*. Crowell, New York.
9. Bartoshuk, L.M. (1978). Gustatory system. In *Handbook of behavioural neurobiology*, Vol. I, *sensory integration* (ed. R.B. Masteron), pp. 503–67. Plenum Press, New York.
10. Land, E. (1977). The retinex theory of color vision. *Scientific American*, **237**, 108–28.
11. Churchland, P.M. and Churchland, P.S. (1990). Could a machine think? *Scientific American*, **262**, 32–7.

12. Churchland, P.S. (1986). *Neurophilosophy: toward a unified understanding of the mind/brain.* MIT Press, Cambridge, Mass.
13. Churchland, P.M. (1989). *A neurocomputational perspective: the nature of mind and the structure of science.* MIT Press, Cambridge, Mass.

CHAPTER SIX

Neural Darwinism : the brain as a selectional system

GERALD M. EDELMAN AND GIULIO TONONI

Our understanding of the biological bases of psychological phenomena has increased considerably during the last several years. The present decade has been declared 'the decade of the brain' in the hope that several major mysteries of the mind will be unravelled by the end of the century through a more extensive exploration of their molecular and cellular counterparts in the nervous system. Underlying many such efforts is the belief that the brain is a particularly complicated computer that implements an intricate program that will be deciphered by neuroscientific studies. According to this view, it should be possible to understand the brain completely by 'reverse engineering', i.e. by looking at its parts and by 'cracking its code'. While recognizing that the study of brain structure and function is centrally important, we think that a true understanding of brain and mind requires the realization that the brain is neither a machine nor the implementation of a computer program. Instead, we consider that the brain is a selectional system (Edelman 1978, 1987, 1989, 1992), in which selection operates upon variation in somatic time, i.e. during the lifetime of an individual. This proposal is based on two main sets of observations:

(1) Individual nervous systems (particularly those of vertebrate species) show enormous structural and functional variability at many levels: molecular, cellular, anatomical, physiological, and behavioural (Edelman 1987). Although there is an obvious commonality of neural structure within a species, the degree of individual variation from brain to brain far exceeds that which could be tolerated for reliable performance in any machine constructed according to current engineering principles. A striking example is provided by the great variety of neurotransmitters and neuropeptides as well as the diversity of receptors, channels, and signalling pathways. Such a profusion,

and the complexity of the possible interactions in space and time, cannot be adequately addressed in terms of functionalist or instructionist theories that emphasize software and consider the structure of the brain as merely constituting sufficient hardware to run software that is cognitively important. We take the contrary view that variation and selection within neural populations play key roles in the development and function of the brain. According to this view, variation is not 'noise' superimposed on a set of programmed procedures but is fundamental to the working of a selectional system. For example, the evolution of a large number of neurotransmitters and neuromodulators would be expected to increase greatly the number of functional circuits that are combinatorially possible within a given anatomical network. A rich pharmacology thus assures the existence of a very rich set of functional network variants providing a basis for selectional events in each individual.

(2) Although instructionist theories require precise input, the world of stimuli encountered by a newborn animal cannot be described adequately as pre-existing, unambiguous information ready to be manipulated according to a set of rules similar to those followed by a computer executing a program. While the real stimulus world certainly obeys the laws of physics, it is not uniquely partitioned into 'objects' and 'events' (Smith and Medin 1981). To survive, an organism must either inherit or create criteria that enable it to partition the world into perceptual categories according to its adaptive needs. Even after that partition occurs as a result of experience, the world remains to some extent an unlabelled place full of novelty (Edelman 1987).

THE THEORY OF NEURONAL GROUP SELECTION

If these observations are correct, an alternative to instructionism is required. The theory of neuronal group selection (TNGS, Edelman 1978) was proposed as an alternative theoretical framework sufficiently broad to connect biology and psychology in a fashion consistent with developmental and evolutionary mechanisms. Like the theories of natural selection and of clonal selection in immunity, the TNGS is a population theory. It argues that the ability of organisms to categorize an unlabelled world and behave in an adaptive fashion arises not from instruction or information transfer but from processes of selection upon variation. Instead of ignoring the observed variance and fluctuations in neuroanatomy and neural dynamics, these are treated as key features that are essential to the function of the nervous system.

The TNGS considers that there is continual generation of diversity in the brain, with selection occurring at various levels. In the embryonic and maturing brain, for example, variation and selection occur in migrating cellular populations and during cell death, as well as during synapse formation; both processes are dramatically reflected in enormous synaptic loss during development (Edelman 1988). In the mature brain, variation and selection are reflected mainly in the differential amplification of synaptic efficacies. This results in neuronal group formation and is a process that is continually modified by re-entrant signalling. To understand neuronal groups and re-entrant signalling requires a closer look at the postulates of the TNGS and at its proper mechanisms.

BASIC POSTULATES OF THE THEORY

The TNGS proposes three mechanisms to account for the production of adaptive behaviour by animals with rich nervous systems (Edelman 1987): *Developmental selection, experimental selection,* and *re-entrant signalling* (Fig. 6.1). Each mechanism acts within and among collectives consisting of hundreds to thousands of strongly interconnected neurones called *neuronal groups*. In addition, the theory proposes that selection through differential synaptic amplification is constrained by the action of evolutionarily derived *value systems*. Key examples of value systems are neuromodulatory systems endowed with diffuse projections that signal the occurrence of events having possible adaptive value for the organism as a whole.

(1) Developmental variation and selection

The structural diversity of the nervous system and the details of neuroanatomy are not strictly programmed by a molecular code. Instead, they arise during development from dynamic epigenetic regulation of cell division, adhesion, migration, death, and neurite extension and retraction (see Changeux and Danchin 1973; Cowan 1978; Edelman 1988; Rakic 1988). Neuronal adhesion and migration are governed by a series of morphoregulatory molecules called CAMs, or cell adhesion molecules, and SAMs, or substrate adhesion molecules (Edelman 1988). These molecules interact at neuronal surfaces and affect the dynamics of cellular interactions as they occur at particular neural sites. The temporal patterns and levels of expression of morphoregulatory molecules, while characteristic of a given anatomical area, are nevertheless dynamically regulated and are subject to epigenetic influences. In certain regions, there is also a large amount of cell death that occurs stochastically in particular developing neuronal populations. Such

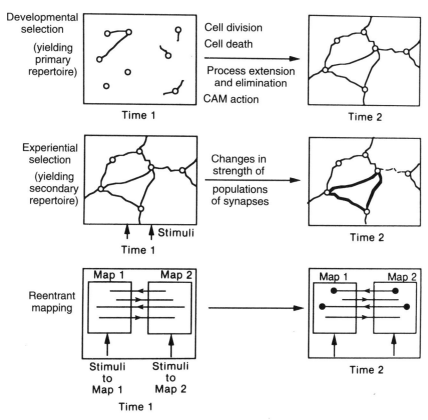

Fig. 6.1. Basic tenets of the theory of neuronal group selection, a global theory of brain function. *Top*: *Developmental selection*. This occurs as a result of the molecular effects of CAM and SAM regulation, growth factor signalling, and selective cell death to yield variant anatomical networks in each individual. These networks make up the *primary repertoire*. *Centre*: *Experimental selection*. Selective strengthening or weakening of populations of synapses as a result of behaviour leads to the formation of various circuits constituting a *secondary repertoire* of neuronal groups. The consequences of synaptic strengthening are indicated by bold paths; of weakening, by dashed paths. *Bottom*: *Re-entry*. Binding of functionally segregated maps occurs in time through parallel selection and temporal correlation of the activities of various maps' neuronal groups. This process provides a fundamental basis for perceptual categorization. Dots at the ends of some of the active reciprocal connections indicate parallel and more-or-less simultaneous strengthening of synapses facilitating certain re-entrant paths. Synaptic strengthening (or weakening) can occur in both the intrinsic and extrinsic re-entrant connections of each map.

processes result in an unavoidable generation of diversity, i.e. they lead to the formation within a given anatomical region of *primary repertoires* containing large numbers of variant neuronal groups or local circuits. This variation in the finest ramifications of neuroanatomical networks occurs despite the fact that the overall pattern in any particular specialized region is similar from individual to individual.

(2) Experiential selection

After most of the anatomical connections of the primary repertoires have been established, the activities of particular functioning neuronal groups continue to be dynamically selected by ongoing mechanisms of synaptic change driven by behaviour and experience. Unlike natural selection in evolution, which results from differential reproduction, experiential selection results from differential amplification of synaptic populations, strengthening some and weakening others without major changes in anatomy. Such synaptic changes do not represent information that is stored in individual connections between single neurones as in connectionist models. Instead, signals act to select variant *populations* of synapses that connect cells within and between neuronal groups. Continued experiential selection leads to the formation of *secondary repertoires* of neuronal groups in response to particular patterns of signals. Because of the changes that occur in synaptic efficacies, upon later encounters with signals of similar types, the previously selected circuits and neuronal groups in such secondary repertoires are more likely to be favoured over others (Fig. 6.1).

(3) Re-entrant signalling

Experiential selection involves statistical signal correlations between groups of pre- and postsynaptic neurones rather than the transmission of coded messages from one neurone to another. Nevertheless, if these statistical correlations are to serve adaptive behaviour, they must reflect the spatiotemporal properties of signals arising in the real world. This is achieved through re-entrant signalling within and among neuronal maps. Neural mappings relating sensory receptor sheets to particular regions of the central nervous system provide one means of enforcing spatiotemporal regularities. In view of the local variance in neural structure and connectivity and the variant statistics of synaptic change, however, specific tags or labels are *not* available to specify any given map position as they are in a computer representation. How then can different maps be coordinated? The TNGS proposes that mapped regions exchange and correlate signals by re-entry (Fig. 6.1). Re-entry can be defined as ongoing parallel signalling between separate neuronal groups occurring along large numbers of ordered anatomical connections in a bidirectional and recursive fashion. Re-entrant signalling can take place via reciprocal connections between and within maps (as seen in cortico–cortical, corticothalamic, and thalamocortical radiations) as well as via more complex arrangements seen in the connections among cortex, basal ganglia, and cerebellum (Edelman 1989). Re-entry is a dynamic process that is inherently parallel and distributed and it should be sharply differentiated from feedback. Feedback is concerned with error correction and defined inputs and outputs, whereas re-entry has no necessary preferred direction and no predefined input or output function. Moreover, unlike re-entry, which involves parallel processes, a given feedback loop, to be effective, involves only a single signal channel or pair of wired connections (Ashby 1966).

Although it can occur within a single map, re-entry usually involves signalling between at least two maps, and it acts through the ordered connections that sample these maps in both space and time. The simultaneous activation of neuronal groups in different maps by a given stimulus and the effects of previous re-entrant activity both tend to strengthen some of the connections between those groups. This coordinated dynamic interaction across maps results in temporal correlation of the responses of a subset of groups to the disjunct signals travelling in separate channels to each map. By these means, distinct operations in different maps that are related to the same perceptual stimulus can be effectively integrated.

A minimal arrangement of two re-entrantly connected maps is called a 'classification couple'. As shown in Fig. 6.1, after multiple encounters with a stimulus, particular patterns of groups are selected in each mapped area. This process is called re-entry because the results of selection are shuttled back and forth between maps in a recursive fashion. With continued re-entrant interactions across maps resulting in temporally coordinated synaptic changes, responses to certain features or attributes of a stimulus object may become strongly correlated. Because of its reciprocal and recursive properties and its parallel structure, re-entry leads to new neuronal responses and it can resolve conflicts arising between the synaptic activities of different mapped areas (Finkel and Edelman 1989).

SOME ESSENTIAL BOUNDARY CONDITIONS

Neuronal groups

Although single neurones could occasionally serve as targets of selection, the TNGS argues that neuronal groups, not single neurones, are the sufficient basis for the mapping interactions and selectional events proposed by the theory. In the presence of numerous densely connected interneurones in regions such as the cortex, it is difficult to imagine a neurone acting as an isolated or individual unit. Neurones within a group are highly intercon- nected, and cooperative local signalling that couples neighbouring neurones into groups would seem to be inevitable. As a consequence, changes in synaptic strengths tend differentially to enhance the adaptive responses of the group as a whole. The collective action of neurones organized in groups would also enhance the reliability of the system, counteracting neuronal death and occasional non-adaptive selectional events affecting individual cells. Most importantly, group organization provides a basic structural underpinning for the process of re-entry which, given the wide lateral spread of overlapping local dendritic and axonal arbours, could hardly be envisaged as occurring specifically from single neurone to single neurone. Finally, group structure and variation is consistent with the events of embryonic development as well as with the plasticity of cortical neuronal function that is found even in the adult. While the structures underlying neuronal groups arise from local anatomical connections, the groups themselves are dynamic entities whose borders and characteristics are affected by such synaptic changes and by the nature of the signals the groups receive.

A great deal of supportive evidence for the existence of neuronal groups has been generated from both anatomical and physiological studies. In regions of the CNS where specific roles can be assigned to neurones, local mosaic arrangements are observed that provide a natural basis for a functional arrangement into neuronal groups. These include ocular dominance columns, blobs, slabs, barrels, fractured somatotopies, etc. The characteristic patchy shape and sparse connectivity of cortical axonal arbours also accords well with the existence of groups. Indeed, electrophysiological experiments in the cat visual cortex have shown that, upon presentation of a stimulus, neighbouring neurones respond in a strongly correlated fashion and thus constitute a neuronal group (Gray and Singer 1989). Moreover, the separate observations of correlations between distant neuronal groups in striate and extrastriate cortex and across the callosum provide further direct evidence for the dependence of long-range selective interactions on the process of re-entry (Engel *et al.* 1991). Cutting such callosal connections abolished these correlations.

Value and value systems

From a selectionist perspective, there are no programs, sets of instructions, or teachers explicitly controlling synaptic changes in neuronal systems (Edelman

1978, 1987). There are, however, structures or constraints in the phenotype that reflect prior evolutionary selection for what we have called 'values' (Reeke *et al.* 1990). *Values* reflect events involving the nervous system that have been selected during evolution because they contribute to adaptive behaviour and to phenotypic fitness. Examples of low-level values are: 'eating is better than not eating' or 'seeing is better than not seeing', as will be illustrated below in connection with Darwin III and Darwin IV. In the absence of specific detailed instructions, evolution has endowed organisms with several means of sensing the occurrence of behaviours having adaptive value and of selecting the neural events that bring them about. Certain specialized structures in the brain, for example the cholinergic and aminergic neuromo-dulatory systems, seem particularly well suited to serve as *value systems*. Much evidence indicates that these neuromodulatory systems possess certain properties of value systems, such as the ability to give a transient but strong response to the occurrence of events having adaptive value, to signal such an occurrence to wide areas of the brain through diffuse projections, and to release substances that modulate changes in synaptic strength. The modula-tion of local synaptic changes by global signals that are associated, directly or indirectly, with evolutionarily selected values, constitutes a major means to effect *value-dependent learning*. According to the TNGS, value-dependent learning is essential in the selection of adaptive behaviours in somatic time.

SYNTHETIC NEURAL MODELLING: IMPLEMENTING AND TESTING THE THEORY

An extraordinary variety of phenomena in present-day neuroscience demands explanation in terms of fundamental principles of brain organization. Although the TNGS has provided unifying insights into several of these phenomena, they are remarkably complicated, interact at many levels, and are difficult to analyse. Insights into their workings can be tested by the use of detailed models and simulations that embody relevant aspects of brain anatomy and physiology. Such models are essential because even a full theoretical description of an animal's brain would not allow one easily to imagine its detailed workings in time and space. Moreover, the multiple layers of control of brain function and the inherent variation and nonlinearity of brain anatomy and physiology frustrate the general or exclusive use of mathematical analysis. For these reasons, we have adopted a different theoretical approach to understanding integrative brain function during behaviour. This approach, called synthetic neural modelling or SNM (Reeke *et al.* 1990), simultaneously correlates large-scale computer simulations of the nervous system, the phenotype, and the changing environment of a designed organism in order to analyse interactions among all three as behaviour develops. SNM enables the interactions of all simulated structural and functional levels, ranging from the molecular to the behavioural, to be analysed as behaviour develops. Given the historical properties of selective systems, this approach has distinct advantages over mathematical analysis alone.

In what follows, we shall briefly describe several such models that illustrate and test the power of the TNGS as applied to different aspects of brain function. These models are concerned with major problems related to neural map plasticity, cortical integration, and the coordination of movement.

Plasticity of cortical maps

One of the significant discoveries of present-day neuroscience is that cortical maps can change and reorganize during experience in a way that is consistent with selectional mechanisms. For instance, in maps of the somatosensory cortex of adult owl and squirrel monkeys, alterations of input due to nerve section or the repeated presentation of particular stimuli lead to changes in individual map boundaries, particularly in areas 3b and 1 (Kaas *et al.* 1983; Merzenich *et al.* 1983). These changes occur both acutely and chronically. Before and after the changes, the maps have receptive fields with sharp continuous borders no thicker than one or two cell diameters, despite the fact that the arbours of input axons from the thalamus extend over much wider distances. A computer simulation (Pearson *et al.* 1987) based on group formation and competition as described by the TNGS shows how these boundaries and their underlying neuronal groups may be dynamically sustained. In each area in which correlated sets of inputs compete for cortical neurones (for example, those from the glabrous or dorsal skin of the hand), the responding neurones appear to segregate into groups that at any one time are non-overlapping and have sharp boundaries. On the other hand, the organization of these simulated groups changes systematically with time under different stimulus conditions, a finding that is consistent with the experimental results.

The problem of cortical integration

The organization of the cerebral cortex is such that, even within a single sensory modality such as vision, there is a multitude of specialized areas and functionally segregated maps. Each of these areas and the groups of neurones within them tend to respond to different specific attributes of object stimuli, such as spatial position, shape, colour, texture, and depth. What we are aware of, however, is a unitary and coherent perceptual scene that is a prerequisite for adaptive behaviour. This poses the central problem of how integration of the activity of neuronal groups distributed across many functionally segregated cortical areas may take place.

The TNGS suggests that the integration is not due to convergence of connections to some hierarchically superordinate 'master area', but rather is a result of a process of re-entrant signalling within and between areas along a system of parallel and reciprocal connections. Within a single cortical area, re-entry results in '*linking*' of the responses of those neuronal groups that belong to the *same* sensory feature domain. Perceptual grouping within a single submodality such as colour or movement provides an example of such integrative linking at an early level. At a higher level, re-entry results in '*binding*' among the responses of neuronal groups found in *different* feature domains distributed in different cortical areas. An example is the integration of neuronal responses to a particular object contour with its colour, position, and direction of movement. Perceptual and behavioural integration with binding across functionally segregated maps can occur in times ranging from 50 to 500 milliseconds, placing strong temporal constraints on any proposed mechanism.

The self-consistency of the proposal that re-entry mediates linking and binding has been tested extensively in a series of computer models (Sporns *et al.* 1989, 1991; Tononi *et al.* 1992). It has been shown that re-entrant interactions within a single cortical area can give rise to temporal correlations between neighbouring as well as distant groups with a near-zero phase lag as experimentally observed in cats and monkeys (Gray and Singer 1989; Engel *et al.* 1991; Kreiter and Singer 1992). An early computer simulation (Sporns *et al.* 1989) showed how linking can be mediated by re-entry. In agreement with the experimental data, when a continuous long bar was presented as a moving stimulus to the model, correlations were found between units in groups with non-overlapping receptive fields. These distinct correlations disappeared if two collinear short bars that were separated by a gap were moved with the same velocity. A more extended model (Sporns *et al.* 1991) was presented with a pattern composed of several bars moving coherently together but embedded in a background of vertical and horizontal bars that were moving at random to the right and left, or up and down. The neuronal groups responding to the bars that moved in the same direction were rapidly linked by coherent oscillations through re-entry, even though the lateral spread of the anatomical connections from each neuronal group was much smaller than the projected size of the 'object'. This model was shown to segregate a figure from another overlapping figure or from a coherent background of identical texture moving in a different direction (Fig. 6.2).

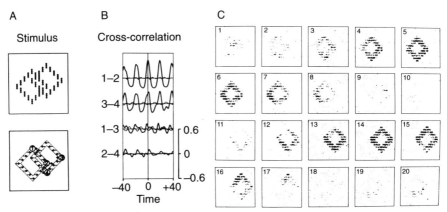

Fig. 6.2. Consequences of re-entry in a neural model of perceptual grouping and segmentation (Sporns *et al.* 1991). (A) A stimulus consisting of two identical patterns each composed of vertically oriented bars is presented to the model. The two patterns overlap in visual space but move in different directions. In the top panel, the bars are shown at their starting positions; in the bottom panel, their corresponding directions of movement are indicated by arrows. Encircled numbers with arrows in the bottom panel refer to the locations of recorded neuronal activity; corresponding cross-correlations are displayed in (B). 'Electrodes' 1 and 2 recorded from neurones responding to Fig. 6.1, and 'electrodes' 3 and 4 from neurones responding to pattern 2. (B) Cross-correlograms computed over a 100-msec sample period and subsequently averaged over 10 trials. Numbers refer to the locations of responding direction-selective repertoires containing neuronal groups that are analysed for their correlations (see A). Four correlograms computed between msec 201 and 300 after stimulus onset are shown. The correlograms are scaled, and shift-predictors (thin lines; averages over 9 shifts) are displayed for comparison. (C) Frames taken from a cine film showing the responses of direction selective groups in the model to the stimulus in (A). The frames show a continuous period of 20 msec (20 iterations) recorded about 150 msec after stimulus onset. Each frame displays the model's entire array of neuronal groups (16 by 16 for a total of 50 720 cells) selective for motion to the right and to the left, and arranged in an interleaved fashion (this accounts for the striped pattern). Each small dot within the array is an active neurone. For the first 10 msec (frames 1–10) groups responding to the pattern moving right are mainly active; subsequently these groups are silent and groups responsive to the other pattern become active (frames 11–20). Note that neuronal activity is strongly correlated both within groups as well as over the entire extent of each pattern. (A and B: modified from Sporns *et al.* 1991, reproduced with permission; C: modified from Tononi *et al.* 1992, reproduced with permission.)

All the simulations that resulted in linking were strongly facilitated by the occurrence of rapid changes in synaptic efficacy. The neural mechanism for integration and segregation of elementary features into objects and background appears to be based on the pattern of temporal correlations and phase relationships among neuronal groups. The correlations were shown to depend critically upon re-entry and disappeared when the underlying connectivity was disrupted. The figural grouping and segregation resulting from re-entry are consistent with the Gestalt laws (Köhler 1947) of continuity, proximity, similarity, common orientation, and common motion, and this work thus suggests a neural basis for these laws.

In a more recent model (Tononi *et al.* 1992), we addressed integration across multiple areas of the visual system, and we coupled the responses of these areas to a simple behavioural output. This approach eliminated the problem of the homunculus, i.e. the need to deduce potential outputs through interpretation of specific patterns of neural activity and correlations from the point of view of a privileged observer. Altogether, ten thousand units were linked by about one million connections between areas at different levels (forward and backward), areas at the same level (lateral), and within a given area (intrinsic) (Fig. 6.3). This model incorporated many important structural features of the organization of the visual system (e.g. Zeki 1981; Felleman and Van Essen 1991). A number of properties were included: different areas deal with different specific characteristics of the stimuli (*functional specialization*). At the same time, they could be grouped into different streams (*parallel organization*) for form, colour, and motion, or into two systems, an 'occipitotemporal' system related to the identity ('what'), and an 'occipito-parietal' system related to the location of an object ('where'). There was a progressive increase in receptive field size from 'lower' to 'higher' areas, and units in these areas acquired more complex response properties (*hierarchical organization*). All areas had systems of horizontal connections which locally and preferentially connect units with similar feature specificity, e.g. similar orientation preference (*intra-areal connections*). Forward *inter-areal connections* were generally organized in such a way that a small region of a 'higher' area received input from a large region of a 'lower' area. This was responsible for the increased size of the receptive fields at 'higher' levels. In general, forward connections tended to preserve or enhance functional specificity, while backward connections were more divergent. In agreement with the experimental literature (see Felleman and Van Essen 1991), all intra-areal and most of the inter-areal pathways in the simulations were reciprocal.

The model contained an *output stage* consisting of a set of oculomotor neurones driving a simulated foveation response that could be used for operant conditioning. In the model, reward was mediated by the activation of a *diffuse*

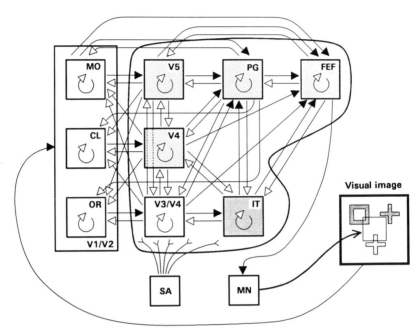

Fig. 6.3. Architecture of the visual cortex model. Segregated visual maps are indicated as boxes; pathways (composed of many thousands of individual connections) are indicated as arrows. Filled arrows indicate voltage-independent pathways, unfilled arrows indicate voltage-dependent pathways. Curved arrows within boxes indicate intra-areal connections. The model comprises three parallel streams involved in the analysis of visual motion (top row), colour (middle row), and form (bottom row). Areas are finely (no shading) or coarsely topographic (light shading), or nontopographic (heavy shading). The visual image (sampled by a colour CCD camera) is indicated at the extreme right. The various areas in the figure are named after areas in the visual cortex that fulfil a similar functional role. For instance, units in V1/V2 OR respond to oriented line segments, units in V3/V4 to corners, and units in IT to entire objects such as crosses or squares in a position-invariant way. Units in V1/V2 CL respond to wavelength, while those in V4 display some degree of colour constancy. Units in V1/V2 MO are responsive to local motion, and those in V5 to pattern motion. The model includes an output stage consisting of area FEF and a set of oculomotor neurones (MN) driving a simulated foveation response, which can be used for operant conditioning. The box labelled SA refers to the diffusely projecting saliency system used in the behavioural paradigm; the general area of projection is outlined. The complete system contains a total of about 10 000 neuronal units and of about 1 000 000 connections.

(From Tononi *et al.* 1992, reproduced with permission.)

projection system or saliency system that was able to affect synaptic changes simultaneously in many cortical areas, thus providing a substrate for selection.

The model's performance illustrates the power of synthetic neural modelling in addressing multiple levels of brain function. To be consistent with functional segregation in the visual cortex, units within each separate area of the model responded to different properties (motion, form, and colour) of the stimuli. Re-entrant interactions within and between the multiple areas gave rise to short-term temporal correlations between the units responding to different attributes of the same stimulus, while units responding to attributes of different stimuli were less correlated. In this way, several features of a single object were 'bound' together across different streams and hierarchical levels, and at the same time that object was differentiated from other objects present in the same visual scene. The resulting specific patterns of activity and correlations often produced a foveation response towards one of the objects. Such response was used as a basis for conditioning the model to foveate an object that was characterized by a certain conjunction of shape and colour, for example a red cross, in any position in the visual field. Conditioning was achieved through the activation, whenever the model foveated the red cross, of diffusely projecting units that signalled saliency (e.g. the delivery of a reward in a conditioning paradigm) and thereby influenced the modification of cortico–cortical connections. After the training period, the model foveated almost exclusively the red cross, although it was presented simultaneously with other objects sharing the same attributes, e.g. a red square and a green cross. The correct response, which depended on the integration of colour, form, and location, was achieved without the need for a hierarchically superordinate area or for units that responded to that particular conjunction. The model thus offered a demonstration that the so-called 'binding problem' can be solved in a way that is fully consistent with known constraints of neuroanatomy and neurophysiology. While the chosen modality in this model was visual, there is every reason to expect that similar principles operate in other modalities throughout the brain. Moreover, the simulations show that the same constructive and correlative effects of re-entry between different areas and streams that offer a solution to the binding problem can provide a basis for understanding several classes of psychophysical effects, for example the construction of form-from-motion boundaries and motion capture.

Complex motor control and the Bernstein problem

Experiments on neural responses of monkeys reaching for targets indicate that a given reaching movement results from the contributions of multiple populations of neurones, each tuned to a particular direction of movement (Georgopoulos *et al.* 1986). A particular movement can thus be explained in terms of the activity of appropriate combinations of neuronal groups. But the development of coordinated movements poses an additional problem that cannot easily be explained by purely cybernetic models of movement control. The presence of multiple degrees of freedom in the joints of the arm and hand indicates that the reaching system is dynamically underdetermined, and therefore that certain constraints are necessary to account for the precise and rapid targeting of movements. This challenging problem of inverse kinematics, first posed by Bernstein (1967), can be solved in terms of the selection of appropriate movements from a repertoire of variant movements that result from underlying neuronal group selection (Sporns and Edelman 1991). This has been explicitly demonstrated in a computer model of a simple simulated organism called Darwin III.

BEHAVING AUTOMATA BASED ON SELECTION

Putting everything together: Darwin III

The three examples of models described above deal with plasticity of sensory maps, perceptual integration, and motor control. The power of SNM can be appreciated in a more complete form by applying it to an entire organism capable of conceptual categorization within a dynamically changing environment. After a behavioural sequence, the entire series of events leading to a given behaviour can be analysed for correlations at all levels ranging from molecular responses to the behaviour itself. Darwin III, perhaps the most sophisticated of the so-called Darwin series of recognition automata (Reeke *et al.* 1990), provides a good example. Darwin III consists of a simple sessile organism with a moveable eye and four-jointed arm. It possesses neurones subserving contrast vision, light touch, kinaesthesia, and motor outputs. A detailed set of neuronanatomical and neurochemical constraints is embedded within its structure, which was designed to embody both evolutionary and developmental steps that might yield such a phenotype.

The environment of Darwin III consisted of objects of different shapes that appeared and moved across its visual field. These objects were chosen and driven by a random number generator. A naïve individual with initial neural activity also driven by random number generators was exposed to these stimuli. After such exposure, selectional events within neuronal repertoires constrained by a value system (Reeke *et al.* 1990) resulted in consistent patterns of visual tracking, reaching with the arm, and discrimination among different objects.

After 'being born' into an environment of moving objects, Darwin III begins to track and fixate particular objects, and to reach out to touch and trace them. With suitable experience leading to synaptic selection, the eye of Darwin III begins to make appropriate saccades and fine tracking movements with no further specification of its task other than that implicit in a value scheme. In a similar fashion, the arm of Darwin III can be trained to reach for and touch objects that are first detected and tracked by the visual system. This performance, which entails the coordination of gestural motions involving various joints (Bernstein 1967), involves participation of a whole series of neural repertoires that perform functions similar to those carried out in real nervous systems. After appropriate experience, the global mapping represented by Darwin III was capable of perceptually segregating striped bumpy 'objects' of various shapes from smooth or non-striped 'objects', despite the fact that its specific behaviour was not driven by an explicit program (Fig. 6.4).

A real world instantiation: Darwin IV

In the latest simulation of the Darwin series, Darwin IV, we have recently applied SNM techniques to a real-world artefact (Edelman *et al.* 1992). This tactic restricts the computer simulation solely to the organization and dynamics of the nervous system; the environment and the artefact are real. Darwin IV thus retains the advantages of SNM while avoiding the difficulties and pitfalls of attempting to simulate a rich environment in addition to a brain. The artefact itself, called NOMAD (Neurally Organized Multiply Adaptive Device), moves about in the enrivonment and provides visual and other sensory inputs to a simulated nervous system (Darwin IV) in a supercomputer. NOMAD in turn receives telemetered signals from the neural portions of Darwin IV that govern its behaviour (Fig. 6.5).

Darwin IV's behavioural repertoire includes several reflex responses (gripping, camera ('eye') elevation, snout sensing, and avoidance) as well as adaptive behaviours resulting from sensorimotor interactions (such as random search, tracking, approaching, and homing) that are subject to selective amplification. These behavioural modes and reflexes are combined during experience to perform a number of exemplary tasks. In general, the individual adaptive behaviours of NOMAD depend upon experience and are not predictable in detail, although they follow constraints imposed by value schemes within Darwin IV.

Darwin IV contains two distinct value systems that were actually wired in but can be thought of as products of evolution. The first can be metaphorically

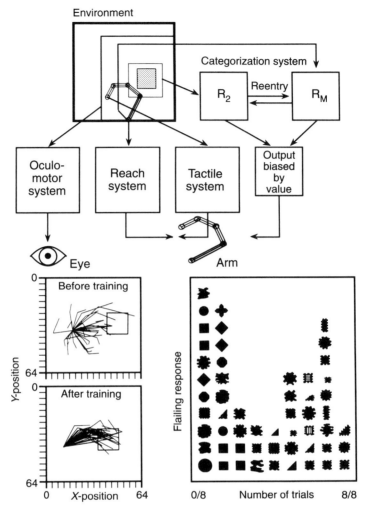

Fig. 6.4. Darwin III, a recognition automaton that performs as a global mapping simulated in a supercomputer. It has a single movable eye, a four-jointed arm with touch at the last joint, and kinaesthesia signalled by neurones in its joints as they move. Its nervous system is organized into several subsystems, each responsible for different aspects of its behaviour (top). What is *programmed* in the simulation is the 'evolutionary' phenotype, including neuroanatomy; the behaviour of the simulation is *not* programmed. After experience with randomly moving objects that it 'sees', its eye will follow any object. Similarly, its arm reaches out to 'touch' an object, and with each selection of movements it increases its success in achieving this touching (lower left). In the experiments shown at the lower left, the tip of the arm always starts in a standard location (the point of origin of the traces) and its motion toward a target area (the square box) is plotted. Notice that before training, the arm moves in many directions. After training involving selection (bottom set of traces), its movements are targeted. Darwin III was confronted (lower right) with fifty-five different objects and was given eight trials in which to categorize each object. The results indicate that Darwin III divided this collection of objects into two classes.

Fig. 6.5. NOMAD. Photograph of NOMAD in its environment as arranged for the block-sorting experiments (showing NOMAD, randomly distributed blocks, and the landmark for 'home position'). NOMAD is based on a battery-powered mobile platform with three steerable wheels that permit independent translational and rotational motion. Modules stacked on the platform provide effectors, sensors, and a telemetry interface to the nervous system of Darwin IV, which is simulated on an nCUBE/10 parallel supercomputer and is based on biologically realistic neuronal networks. A rigid 'snout' fitted with an electromagnet permits NOMAD to grip small metal objects. NOMAD's primary visual input is provided by a miniature colour CCD video camera mounted near the top of the device. Camera azimuth can be varied by rotation of the platform as a whole; elevation can be directly controlled by NOMAD's nervous system. Additional 'senses' are provided by electrical contacts in the snout which can detect the conductivity of objects gripped by the electromagnet ('taste'), as well as by infrared proximity sensors mounted round the periphery of the base. Experiments with NOMAD are carried out in a specially arranged room, equipped with a frame containing an 8 ft × 10 ft raised floor and surrounded by 'walls' consisting of projection screens. Depending on the studies to be carried out, the experimental area may be configured to include a collection of objects and can contain real or projected landmarks upon the floor or walls. The area under the raised floor contains a second mobile platform controlled by a conventional robotics program. This platform carries a flashlight that projects a stimulus upward through the translucent floor for tracking by NOMAD. A television camera mounted on the ceiling is used to record the behaviour of NOMAD for later evaluation.

described as specifying that 'seeing is better than not seeing'. In the networks of Darwin IV it is implemented by visual units that respond more strongly when a target appears in the region of the visual field adjacent to NOMAD's snout. Its activity influences the probability of occurrence of changes in synaptic strength between visual and motor networks leading to tracking. This means that when the central retina becomes strongly stimulated, the strengths of those synapses that were active in the recent past and thus potentially involved in the behaviour that brought about the increased stimulation are selectively increased through value-dependent modulation. Note that Darwin IV contains no programmed specification of *how* light-tracking movements are to be accomplished.

The second value system is triggered whenever Darwin IV activates its snout sensor to assess surface conductivity. Value-dependent modulation of synaptic strengths occurs in the connections linking visual repertoires with reflex centres. As described below, this results in NOMAD's ability to collect blocks of a certain colour and actively to avoid blocks of a different colour. It should be emphasized that, since the resulting synaptic changes in the value-dependent system are probabilistic, no two versions of Darwin IV show identical behaviour. Nevertheless, their behaviour tends to converge in directions favoured by their intrinsic values, and by the phenotype.

The capabilities of Darwin IV have been tested in two real-world tasks: light tracking and block sorting.

Light tracking. As a first example of a simple behaviour, we investigated the ability of Darwin IV to approach and track a light spot moving along random trajectories projected from below on the translucent floor of the experimental room. This task is a real-world extension of the tracking task already successfully addressed by Darwin III. NOMAD's video camera was installed at the front of the device pointing at an angle of roughly 45° down from the horizontal. In the absence of a visual stimulus, NOMAD rotates and translates at random, driven by spontaneous activity in its motor repertoires. When the light appears in the visual field, the motor repertoires of Darwin IV are directly activated by the visual network. After a movement occurs, the synaptic populations giving rise to this movement are probabilistically strengthened or weakened by selection, depending on whether or not the movement resulted in an increase in value. After some time, those movements that facilitate close approaches and tracking occur more frequently than others.

During early trials, NOMAD rarely approaches the light. However, even if NOMAD approaches the light by chance, it is unable to track the target's movements because there is no innate adaptation to carry out this task. After experience and the concomitant selectional events in its nervous system,

NOMAD consistently tracks the light along complicated trajectories and loses contact only occasionally. When contact is lost, NOMAD briefly reverts to random movements, but it resumes tracking after encountering the light again. We have noted that tracking ability was acquired by Darwin IV using the same selectional scheme that had already been employed in Darwin III with minor modifications; the time required to learn tracking was comparable. This indicates the generality of selectional mechanisms: these work well within limits even when many important parameters change. For example, in Darwin IV the oblique angle of the camera gives rise to 'keystone' distortion of the visual image and distant stimuli appear to grow larger when approached. Despite these changes, selective modification of synaptic strengths during training automatically accommodates to any nonlinearities resulting from the distorted visual image and from the mechanics of NOMAD's motor apparatus.

Object sorting. Following successful training for the tracking task, we used the behavioural gains developed by Darwin IV while tracking a moving light for a different but related task: locating and approaching stationary objects (Fig. 6.5). We devised a simple block sorting task using conductive blue blocks and nonconductive red blocks. The aim of this task was to train Darwin IV, through experience with 'good- and bad-tasting' blocks, to avoid those blocks whose colour is associated with 'bad taste' without the need to approach and grip them. Before training, NOMAD approached all blocks in its vicinity under the guidance of its tracking system, and, in approximately 50 per cent of all approaches, it established physical contact with the surface of a target block. In these cases, the block was gripped with the magnetic snout and the camera was raised by reflex into a horizontal position. The resulting change in Darwin IV's field of view allowed it to use the tracking system to search for a designated home position while pushing the block with its snout across the floor. As soon as the home position was reached, the camera turned downward again by reflex towards the block and the conductive properties of the block were sensed with the snout, leading to a strong signal in the value system. This signal increased the probability of concurrent synaptic changes within Darwin IV's nervous system regardless of whether the block was in fact conductive or nonconductive. In general, the result of such activity is a strengthening of the association between the active vision repertoires and the gripping reflex. As a consequence, the probability of gripping an object in the future increases. If high conductivity ('bad taste') is sensed, however, an avoidance reflex is activated; this causes NOMAD to turn away from the direction of the object. Inherited (i.e. evolutionarily selected) value thus allows an association to be formed between the colour of the object in the foveal region and an aversive response. Because of the resulting changes in synaptic strengths, in future encounters the appearance in the perifoveal region of an object of the colour previously sensed as having 'bad taste' is sufficient to trigger an avoidance response. After having delivered an object of either colour to the home position, NOMAD resumes its search for new objects after habituation in the visual networks sets in. The result of the search is an accumulation of red ('good-tasting') blocks at 'home' and an avoidance of blue ('bad-tasting') blocks.

These two examples of Darwin IV's performance demonstrate some important insights emerging from the use of SNM based on a selectional theory. Some of the tasks, for example block sorting, require the experience-dependent combination of sequences of reflex and adaptive behaviours. This behavioural combination occurs as a result of neuronal group selection based on value, and when value circuits are cut or interfered with the behaviour does not occur. Each behaviour is controlled by multiple repertoires in the nervous system; a strict one-to-one mapping does not exist between neural centres and the behaviours elicited by their activation. Darwin IV provides a valuable testing ground for neural models in real-world environments. It allows the investigator to explore the effects of many possible combinations of phenotypic structure and neuronal architecture, and to test certain psychological implications of the TNGS, particularly those related to learning. A full review of its behavioural performance can be carried out in terms of all its levels of organization—neuroanatomical, physiological, and behavioural—as well as their statistical correlations. The results provide a heuristically valuable basis for evaluating experiments carried out in animals to test the theory.

CONCLUSION

In this chapter we have argued that understanding how adaptive behaviour in a rich environment can be initiated and controlled by vertebrate nervous systems requires the development and testing of global theories of brain function. The TNGS is such a theory; its main assumptions have been described and illustrated here with a few examples taken both from neurobiology and from computer simulations. The use of SNM, although still at a very early stage, has allowed us to explore certain important psychological questions in terms of the basic biological components of which nervous systems are constructed. The results have provided a glimpse of how we might eventually come to understand the means by which behaviour, learning, memory, and even consciousness appear as consequences of the workings of selective mechanisms in the brain. The interested reader may consult several publications (Edelman 1987, 1989, 1992) which describe how, without further assumptions, the theory can account for key features of such higher-order brain functions.

The philosophical question of whether real brains can be reduced to and eventually understood as machines or computers implementing a particular program is much more than just a matter of vocabulary. The adoption of such a view would require mechanisms for the genetic specification of detailed neural codes, for the unambiguous definition of categories, and for the generation and replication of sequences of instructions that are essentially incompatible with what we currently know about the nature and development of biological systems. In contrast, according to the selectionist approach taken by the TNGS, many aspects of individual history necessarily enter into each and every neuronal response. It is such a selectional history constrained by evolutionarily derived values that endows us with the unpredictability and individuality that makes us human.

Acknowledgement

The work performed at the Neurosciences Institute cited here was supported by the Neurosciences Research Foundation.

REFERENCES

Ashby, W.B. (1966). *An introduction to cybernetics*. Wiley, New York.

Bernstein, N. (1967). *The coordination and regulation of movements*. Pergamon Press, Oxford.

Changeux, J.P. and Danchin A. (1973). Selective stabilization of developing synapses as a mechanism for the specification of neuronal networks. *Nature*, **264**, 705–11.

Cowan, W.M. (1978). Aspects of neural development. *International Review of Physiology*, **17**, 150–91.

Edelman, G.M. (1978). Group selection and phasic re-entrant signalling: A theory of higher brain function. In *The mindful brain* (ed. G.M. Edelman and V.B. Mountcastle). MIT Press, Cambridge, Mass.

Edelman, G.M. (1987). *Neural Darwinism: the theory of neuronal group selection*. Basic Books, New York.

Edelman, G.M. (1988). *Topobiology: an introduction to molecular embryology*. Basic Books, New York.

Edelman, G.M. (1989). *The remembered present : a biological theory of consciousness*. Basic Books, New York.

Edelman, G.M. (1992). *Bright air, brilliant fire : on the matter of the mind*. Basic Books, New York.

Edelman, G.M., Reeke, G.N. Jr, Gall, W.E., Tononi, G., Williams, D., and Sporns, O. (1992). Synthetic neural modeling applied to a real-world artifact. *Proceedings of the National Academy of Sciences of the USA*, **89**, 7267–71.

Engel, A.K., König, P., Kreiter, A.K. and Singer, W. (1991). Interhemispheric synchronization of scillatory neuronal responses in cat visual cortex. *Science*, **252**, 1177–9.

Felleman, D.J. and Van Essen, D. C. (1991). Distributed hierarchical processing in the primate cerebral cortex. *Cerebral Cortex*, **1**, 1–47.

Finkel, L.H. and Edelman, G.M. (1989). The integration of distributed cortical systems by re-entry: a computer simulation of interactive functionally segregated visual areas. *Journal of Neuroscience*, **9**, 3188–208.

Georgopoulos, A.P., Schwartz, A.B., and Ketner, R.E. (1986). Neuronal population coding of movement direction. *Science*, **233**, 1416–19.

Gray, C.M. and Singer, W. (1989). Stimulus-specific neuronal oscillations in orientation columns of cat visual cortex. *Proceedings of the National Academy of Sciences of the USA*, **86**, 1698–702.

Kaas, J.H., Merzenich, M.M., and Killackey, H.P. (1983). The reorganization of somatosensory cortex following peripheral-nerve damage in adult and developing mammals. *Annual Review of Neuroscience*, **6**, 325–56.

Köhler, W. (1947). *Gestalt psychology*. Liverwright, New York.

Kreiter, A.K. and Singer, W. (1992). Oscillatory neuronal response in the visual cortex of the awake macaque monkey. *European Journal of Neurosciences*, **4**, 369–75.

Merzenich, M.M., Kaas, J.H., Wall, J.T., Nelson, T.J., Sur, M., and Felleman, D.J. (1983). Topographic reorganization of somatosensory cortical areas 3b and 1 in adult monkeys following restricted deafferentation. *Neuroscience*, **8**, 33–55.

Pearson, J.C., Finkel, L. H., and Edelman, G.M. (1987). Plasticity in the organization of adult cortical maps: a computer simulation based on neuronal group selection. *Journal of Neuroscience*, **7(12)**, 4209–23.

Rakic, P. (1988). Specification of cerebral cortical areas. *Science*, **241**, 170–6.

Reeke, G.N. Jr, Finkel, L.H., Sporns, O., and Edelman, G.M. (1990). Synthetic neural modelling: A multilevel approach to the analysis of brain complexity. In *Signal and sense: local and global order in perceptual maps* (ed. G.M. Edelman, W.E. Gall, and W.M. Cowan), pp. 607–707. Wiley, New York.

Smith, E.E. and Medin, D.L. (1981). *Categories and concepts*. Harvard University Press, Cambridge, Mass.

Sporns, O. and Edelman, G.M. (1991). Solving Berstein's problem: a proposal for the development of coordinated movement by selection. *Child Development*, **64**, 960–81.

Sporns, O., Gally, J.A., Reeke, G.N. Jr, and Edelman, G.M. (1989). Re-entrant signalling among simulated neuronal groups leads to coherency in their oscillatory activity. *Proceedings of the National Academy of Sciences of the USA*, **86**, 7265–9.

Sporns, O., Tononi, G., and Edelman, G.M. (1991). Modelling perceptual grouping and figure-ground segregation by means of active re-entrant connections. *Proceedings of the National Academy of Sciences of the USA*, **88**, 129–33.

Tononi, G., Sporns, O., and Edelman, G.M. (1992). Re-entry and the problem of integrating multiple cortical areas: simulation of dynamic integration in the visual system. *Cerebral Cortex*, **2**, 310–35.

Zeki, S. (1981). The mapping of visual functions in the cerebral cortex. In *Brain mechanisms of sensation: the Third Taniguchi Symposium on Brain Sciences* (ed. Y. Katsuki, R. Norgren, and M. Sato), pp. 105–28. Wiley, New York.

CHAPTER SEVEN

A new vision of the mind

OLIVER SACKS

1

Five years ago the concepts of 'mind' and 'consciousness' were virtually excluded from scientific discourse. Now they have come back, and every week we see the publication of new books on the subject. Reading most of this work, we may have a sense of disappointment, even outrage; beneath the enthusiasm about scientific developments, there is a certain thinness, a poverty and unreality compared to what we know of human nature, the complexity and density of the emotions we feel and of the thoughts we have. We read excitedly of the latest chemical, computational, or quantum theory of mind, and then ask, 'Is that all there is to it?'

I remember the excitement with which I read Norbert Wiener's *Cybernetics* when it came out in the late 1940s. And then, in the early 1950s, reading the work of Wiener's younger colleagues at MIT—a galaxy of some of the finest minds in America, including Warren McCulloch, Walter Pitts, and John von Neumann; reading about their pioneer explorations of logical automata and nerve nets, I thought, as many of us did, that we were on the verge of computer translation, perception, cognition; a brave new world in which ever more powerful computers would be able to mimic, and even take over, the chief functions of brain and mind. The very titles of the MIT papers were exalted and thrilling: 'machines that think and want', 'The genesis of social evolution in the mindlike behaviour of artifacts'.[1]

During the 1960s, there was some faltering and questioning: it proved possible to put a man on the moon in this decade, but not possible for a computer to achieve a decent translation of a child's speech, much less a text of any complexity, or to achieve more than the most rudimentary mechanical perception (if indeed 'perception' was a legitimate word here).[2] Or was it simply that one needed more computer power, and perhaps different programs or designs? Supercomputers emerged, and, soon, so-called neural networks, which do not consist of actual neurones but computer simulations or models that attempt to mimic the nervous system. Though such networks start with random connections, and learn in a fashion—for example, how to recognize faces or words—they are always instructed what to do, even if they are not instructed how to do it. They are able to recognize in a formal, rulebound way, but not in terms of context and meaning, the way an organism does.

Some of these networks have been developed on the West Coast, under the presiding genius of Francis Crick. And yet Crick himself has expressed fundamental reservations about them: can they, he has asked, really be said to think? Are they, in fact, like minds at all? We must indeed be very cautious before we allow that any artefact is (except in a superficial sense) 'mind-like' or 'brainlike'.[3]

Thus if we are to have a model or theory of mind as this actually occurs in living creatures in the world, it may have to be radically different from anything like a computational one. It will have to be grounded in biological reality, in the anatomical and developmental and functional details of the nervous system; and also in the inner life or mental life of the living creature, the play of its sensations and feelings and drives and intentions, its perception of objects and people and situations, and, in higher creatures at least, the ability to think abstractly and to share through language and culture the consciousness of others.

Above all such a theory must account for the development and adaptation peculiar to living systems. Living organisms are born into a world of challenge and novelty, a world of significances, to which they must adapt or die. Living organisms grow, learn, develop, organize knowledge, and use memory—in a way that has no real analogue in the non-living.[4] Memory itself is characteristic of life. And memory brings about a change in the organism, so that it is better adapted, better fitted, to meet environmental challenges. The very 'self' of the organism is enlarged by memory.

Such a notion of organic change as taking place with experience and learning, and as being an essential change in the structure and 'being' of the organism, had no place in the classical theories of memory, which tended to portray it as a thing-in-itself, something *deposited* in the brain and mind: an impression, a trace, a replica of the original experience, like a photograph. (For Socrates, the brain was soft wax, imprinted with impressions as with a seal or signet ring.) This was certainly the case with Locke and the empiricists, and has its counterpart in many of the current models of memory, which see it as having a definite location in the brain, something like the memory core of a computer.

The neural basis of memory, and of learning generally, the psychologist

Donald Hebb hypothesized, lay in a selective strengthening or inhibition of the synapses between nerve cells and the development of groups of cells or 'cell-assemblies' embodying the remembered experience. This change, for Hebb, was only a local one, not a change in the brain (or the self) as a whole. At the opposite extreme, his teacher Karl Lashley, who trained rats to do complex tasks after removing various parts of their brains, came to feel that it was impossible to localize memory or learning; that, with remembering and learning, changes took place throughout the entire brain. Thus, for Lashley, memory, and indeed identity, did not have discrete locations in the brain.[5] There seemed no possible meeting-point between these two views: an atomistic or mosaic view of the brain as parcelling memory and perception into small, discrete areas, and a global or 'gestalt' view, which saw them as being somehow spread out across the entire brain.

These disparate views of memory and brain function were only part of a more general chaos, a flourishing of many fields and many theories, independently and in isolation, a fragmentation of our approaches to, and views about, the brain. A comprehensive theory of brain function that could make sense of the diverse observations of a dozen different disciplines has been missing, and the enormous but fragmented growth of neuroscience in the last two decades has made the need for such a general theory more and more pressing. This was well expressed in a 1992 article in *Nature*, in which Jeffrey Gray spoke of the tendency of neuroscience to gather more and more experimental data, while lacking 'a new theory . . . that will render the relations between brain events and conscious experience "transparent"'.[6]

The needed theory, indeed, must do more: it must account for (or at least be compatible with) all the facts of evolution and neural development and neurophysiology, on the one hand, and all the facts of neurology and psychology, on the other. It must be a theory of self-organization and emergent order at every level and scale, from the scurrying of molecules and their micropatterns in a million synaptic clefts to the grand macropatterns of an actual lived life. Such a theory, Gray feels, 'is at present unimaginable'.

But just such a theory has been imagined, and with great force and originality, by Gerald Edelman, who, with his colleagues at the Neurosciences Institute at Rockefeller University over the past fifteen years, has been developing a biological theory of mind, which he calls neural Darwinism, or the Theory of Neuronal Group Selection (TNGS).[7]

Edelman's early work dealt not with the nervous system, but with the immune system, by which all vertebrates defend themselves against invading bacteria and viruses. It was previously accepted that the immune system 'learned', or was 'instructed', by means of a single type of antibody which moulded itself round the foreign body, or antigen, to produce an appropriate, 'tailored' antibody. These moulds then multiplied and entered the bloodstream and destroyed the alien organisms. But Edelman showed that a radically different mechanism was at work; that we possess not one basic kind of antibody, but millions of them, an enormous repertoire of antibodies, from which the invading antigen 'selects' one that fits. It is such a selection, rather than a direct shaping or instruction, that leads to the multiplication of the appropriate antibody and the destruction of the invader. Such a mechanism, which he called a 'clonal selection', was suggested in 1959 by MacFarlane Burnet, but Edelman was the first to demonstrate that such a selectional mechanism actually occurs, and for this he shared a Nobel Prize in 1972.

Edelman then began to study the nervous system, to see whether this too was a selectional system, and whether its workings could be understood as evolving, or emerging, by a similar process of selection. Both the immune system and the nervous system can be seen as systems for recognition. The immune system has to recognize all foreign intruders, to categorize them, reliably, as 'self' or 'not self'. The task of the nervous system is roughly analogous, but far more demanding: it has to classify, to categorize, the whole sensory experience of life, to build from the first categorizations, by degrees, an adequate model of the world; and in the absence of any specific programming or instruction to discover or create its own way of doing this. How does an animal come to recognize and deal with the novel situations it confronts? How is such individual development possible?

The answer, Edelman proposes, is that an evolutionary process takes place—not one that selects organisms and takes millions of years, but one that occurs within each particular organism and during its lifetime, by competition among cells, or selection of cells (or, rather, cell groups) in the brain. This for Edelman is 'somatic selection'.

Edelman and his colleagues have been concerned not only to propose a principle of selection but to explore the mechanisms by which it may take place. Thus they have tried to answer three kinds of questions. Which units in the nervous system select and give different emphasis to sensory experience? How does selection occur? What is the relation of the selecting mechanisms to such functions of brain and mind as perception, categorization, and, finally, consciousness?

Edelman discusses two kinds of selection in the evolution of the nervous system: 'developmental' and 'experiential'. The first takes place largely before birth. The genetic instructions in each organism provide general constraints for neural development, but they cannot specify the exact destination of each developing nerve cell, for these grow and die, migrate in great numbers and in entirely unpredictable ways; all of them are 'gypsies', as Edelman likes to say. Thus the vicissitudes of fetal development themselves produce in every brain unique patterns of neurones and neuronal groups ('developmental selection'). Even identical twins with identical genes will not have identical brains at birth: the fine details of cortical circuitry will be quite different. Such variability, Edelman points out, would be a catastrophe in virtually any mechanical or computational system, where exactness and reproducibility are of the essence. But in a system in which selection is central, the consequences are entirely different; here variation and diversity are themselves of the essence.

Now, already possessing a unique and individual pattern of neuronal groups through developmental selection, the creature is born, thrown into the world, there to be exposed to a new form of selection based upon experience ('experiential selection'). What is the world of a newborn infant (or chimp) like? Is it a sudden incomprehensible (perhaps terrifying) explosion of electromagnetic radiations, sound waves, and chemical stimuli which make the infant cry and sneeze? Or an ordered, intelligible world, in which the infant discerns people, objects, events, and scenes?[8] We know that the world encountered is not one of complete meaninglessness and pandemonium, for the infant shows selective attention and preferences from the start.

Clearly there are some innate biases or dispositions at work; otherwise the infant would have no tendencies whatever, would not be moved to do anything, seek anything, to stay alive. These basic biases Edelman calls 'values'. Such values are essential for adaptation and survival; some have been developed through aeons of evolution; and some are acquired through exploration and experience. Thus if the infant instinctively values food, warmth, and contact with other people (for example), this will direct its first movements and strivings. These 'values'—drives, instincts, intentionalities— serve to weight experience differentially, to orient the organism toward survival and adaptation, to allow what Edelman calls 'categorization *on* value' (e.g. to form categories such as 'edible' and 'nonedible' as part of the process of getting food). It needs to be stressed that 'values' are experienced, internally, as feelings: without feeling there can be no animal life. 'Thus', in the words of the late philosopher Hans Jonas, 'the capacity for feeling, which arose in all organisms, is the mother-value of all values'.

At a more elementary physiological level, there are various sensory and motor 'givens', from the reflexes that automatically occur (for example, in response to pain) to innate mechanisms in the brain, as, for example, the feature detectors in the visual cortex that, as soon as they are activated, detect verticals, horizontals, boundaries, angles, etc., in the visual world.

Thus we have a certain amount of basic equipment; but, in Edelman's view, very little else is programmed or built-in.[9] It is up to the infant animal, given its elementary physiological capacities, and given its inborn values, to create its own categories and to use them to make sense of, to *construct* a world—and it is not just a world that the infant constructs, but its own world, a world constituted from the first by personal meaning and reference.

Such a neuro-evolutionary view is highly consistent with some of the conclusions of psychoanalysis and developmental psychology—in particular, the psychoanalyst Daniel Stern's description of 'an emergent self', 'Infants seek sensory stimulation', writes Stern. 'They have distinct biases or preferences with regard to the sensations they seek . . . . These are innate. From birth on, there appears to be a central tendency to form and test hypotheses about what is occurring in the world . . . [to] categorize . . . into conforming and contrasting patterns, events, sets, and experiences.'[10] Stern emphasizes how crucial are the active processes of connecting, correlating, and categorizing information, and how with these a distinctive organization emerges, which is experienced by the infant as the sense of a self.

It is precisely such processes that Edelman is concerned with. He sees them as grounded in a process of selection acting upon the primary neuronal units with which each of us is equipped. These units are not individual nerve cells or neurones, but groups ranging in size from about fifty to ten thousand neurones; there are perhaps a hundred million such groups in the entire brain. During the development of the fetus, a unique neuronal pattern of connections is created and then, in the infant, experience acts upon this pattern, modifying it by selectively strengthening or weakening connections between neuronal groups, or creating entirely new connections.

Thus experience itself is not passive, a matter of 'impressions' or 'sense-data', but active, and constructed by the organism from the start. Active experience 'selects', or carves out, a new, more complexly connected pattern of neuronal groups, a neuronal reflection of the individual experience of the child, of the procedures by which it has come to categorize reality.

But these neuronal circuits are still at a low level. How do they connect with the inner life, the mind, the behaviour of the creature? It is at this point that Edelman introduces the most radical of his concepts: the concepts of 'maps' and 're-entrant signalling'. A 'map', as he uses the term, is not a representation in the ordinary sense, but an interconnected series of neuronal groups that responds selectively to certain elemental categories: for example, to movements or colours in the visual world. The creation of maps, Edelman postulates, involves the synchronization of hundreds of neuronal groups. Some mappings, some categorizations, take place in discrete and anatomically fixed (or 'prededicated') parts of the cerebral cortex: thus colour is 'constructed' in an area called V4. The visual system alone, for example, has over thirty different maps for representing colour, movement, shape, etc.

But where perception of *objects* is concerned, the world, Edelman likes to say, is not 'labelled', it does not come 'already parsed into objects'. We must *make* them, in effect, through our own categorizations: 'Perception makes', Emerson said. 'Every perception', says Edelman, echoing Emerson, 'is an act of creation'. Thus our sense organs, as we move about, take samplings of the world, creating maps in the brain. Then a sort of neurological 'survival of the fittest' occurs, a selective strengthening of those mappings that correspond to 'successful' perceptions—successful in that they prove the most useful and powerful for the building of 'reality'.

In this view, there are no innate mechanisms for complex 'personal' recognition, such as the 'grandmother cell' postulated by researchers in the 1970s to correspond to one's perception of one's grandmother.[11] Nor is there any 'master area', or 'final common path', whereby all perceptions relating (say) to one's grandmother converge in one single place. There is no such place in the brain where a final image is synthesized, nor any miniature person or homunculus to view this image. Such images or representations do not exist in Edelman's theory, nor do any such homunculi. (Classical theory, with its concept of 'images' or 'representations' in the brain, demanded a sort of dualism, for there had to be a miniature 'someone in the brain' to view the images; and then another, still smaller, someone in the brain of that someone; and so on, in an infinite regress. There is no way of escaping from this regress, except by eliminating the very concept of images and viewers, and replacing it by a dynamic concept of process or interaction.)

Rather, the perception of a grandmother or, say, of a chair, depends on the synchronization of a number of scattered mappings throughout the visual cortex: mappings relating to many different perceptual aspects of the chair (its size, its shape, its colour, its 'leggedness', its relation to other sorts of chairs—armchairs, kneeling chairs, baby chairs, etc.); and perhaps in other parts of the cortex as well (relating to the feel of sitting in a chair, the actions needed to do it, etc.). In this way the brain, the creature, achieves a rich and flexible percept of 'chairhood', which allows the recognition of innumerable sorts of chairs as chairs (computers, by contrast, with their need for unambiguous definitions and criteria, are quite unable to achieve this). This perceptual generalization is dynamic and not static, and depends on the active and incessant orchestration of countless details. Such a correlation is possible because of the very rich connections between the brain's map connections, which are reciprocal, and may contain millions of fibres.

These extensive connections allow what Edelman calls 're-entrant signalling', a continuous 'communication' *between* the active maps themselves, which enables a coherent construct such as 'chair' to be made. This construct arises from the interaction of many sources. Stimuli from, say, touching a chair may affect one set of maps, stimuli from seeing it may affect another set.

Re-entrant signalling takes place between the two sets of maps and between many other maps as well, as part of the process of perceiving a chair.

This construct, it must be emphasized once again, is not comparable to a single image or representation. It is, rather, comparable to a giant and continually modulating equation, as the outputs of innumerable maps, connected by re-entry, not only complement one another at a perceptual level but are built up to higher and higher levels. For the brain, in Edelman's vision, makes maps of its own maps, or 'categorizes its own categorizations', and does so by a process that can ascend indefinitely to yield ever more generalized pictures of the world.

This re-entrant signalling is different from the process of 'feedback', which merely corrects errors.[12] Simple feedback loops are not only common in the technological world (as thermostats, governors, cruise controls, etc.) but are crucial in the nervous system, where they are used for control of all the body's automatic functions from temperature to blood pressure to the fine control of movement. (This concept of feedback is at the heart of both Wiener's cybernetics and Claude Bernard's concept of homeostasis.) But at higher levels, where flexibility and individuality are all-important, and where new powers and new functions are needed and created, one requires a mechanism that can construct, not just control or correct.

The process of re-entrant signalling, with its thousands or hundreds of thousands of reciprocal connections within and between maps, may be likened to a sort of neural United Nations, in which dozens of voices are talking together, while including in their conversation a variety of constantly inflowing reports from the outside world, and giving them coherence, bringing them together into a larger picture as new information is correlated and new insights emerge. There is, to continue the metaphor, no secretary-general in the brain; the activity of re-entrant signalling itself achieves the synthesis. How is this possible?

Edelman, who himself once planned to be a concert violinist, uses musical metaphors here. 'Think', he said in a 1993 BBC radio broadcast, 'if you had a hundred thousand wires randomly connecting four string quartet players and that, even though they weren't speaking words, signals were going back and forth in all kinds of hidden ways [as you usually get them by the subtle nonverbal interactions between the players] that make the whole set of sounds a unified ensemble. That's how the maps of the brain work by re-entry.'

The players are connected. Each player, interpreting the music individually, constantly modulates and is modulated by the others. There is no final or 'master' interpretation: the music is collectively created. This, then, is Edelman's picture of the brain, an orchestra, an ensemble—but without a conductor, an orchestra which makes its own music.

The construction of perceptual categorizations and maps, the capacity for generalization made possible by re-entrant signalling, is the beginning of psychic development, and far precedes the development of consciousness or mind, or of attention or concept formation—yet it is a prerequisite for all of these; it is the beginning of an enormous upward path, and it can achieve remarkable power even in relatively primitive animals like birds.[13] Perceptual categorization, whether of colours, movements, or shapes, is the first step, and it is crucial for learning, but it is not something fixed, something that occurs once and for all. On the contrary—and this is central to the dynamic picture presented by Edelman—there is then a continual recategorization, and this itself constitutes memory.

'In computers', Edelman writes, 'memory depends on the specification and storage of bits of coded information'. This is not the case in the nervous system. Memory in living organisms, by contrast, takes place through activity and continual recategorization.

By its nature, memory . . . involves continual motor activity . . . in different contexts. Because of the new associations arising in these contexts, because of changing inputs and stimuli, and because different combinations of neuronal groups can give rise to a similar output, a given categorical response in memory may be achieved in several ways. Unlike computer-based memory, brain-based memory is inexact, but it is also capable of great degrees of generalization.

2

In the extended theory of neuronal group selection, which he has developed since 1987, Edelman has been able, in a very economical way, to accommodate all the 'higher' aspects of mind—concept formation, language, consciousness itself—without bringing in any additional considerations. Edelman's most ambitious project, indeed, is to try to delineate a possible biological basis for consciousness. He distinguishes, first, 'primary' from 'higher-order' consciousness:

Primary consciousness is the state of being mentally aware of things in the world, of having mental images in the present. But it is not accompanied by any sense of [being] a person with a past and a future . . . In contrast, higher-order consciousness involves the recognition by a thinking subject of his or her own acts and affections. It embodies a model of the personal, and of the past and future as well as the present . . . It is what we as humans have in addition to primary consciousness.

The essential achievement of primary consciousness, as Edelman sees it, is to bring together the many categorizations involved in perception into a *scene*. The advantage of this is that 'events that may have had significance to an animal's past learning can be related to new events'. The relation established will not be a causal one, one necessarily related to anything in the outside world; it will be an *individual* (or 'subjective') one, based on what has had 'value' or 'meaning' for the animal in the past.

Edelman proposes that the ability to create scenes in the mind depends upon the emergence of a new neuronal circuit during evolution, a circuit allowing for continual re-entrant signalling between, on the one hand, the parts of the brain where memory of such value categories as warmth, food, and light takes place and, on the other, the ongoing global mappings that categorize perceptions as they actually take place. This 'bootstrapping process' (as Edelman calls it) goes on in all the senses, thus allowing for the construction of a complex scene. The 'scene', one must stress, is not an image, not a picture (any more than a 'map' is), but a correlation between different kinds of categorization.

Mammals, birds, and some reptiles, Edelman speculates, have such a scene-creating primary consciousness; and such consciousness is 'efficacious'; it helps the animal adapt to complex environments. Without such consciousness, life is lived at a much lower level, with far less ability to learn and adapt.

Primary consciousness [Edelman concludes] is required for the evolution of higher-order consciousness. But it is limited to a small memorial interval around a time chink I call the present. It lacks an explicit *notion* or a concept of a personal self, and it does not afford the ability to model the past or the future as part of a correlated scene. An animal with primary consciousness sees the room the way a beam of light illuminates it. Only that which is in the beam is explicitly in the remembered present; all else is darkness. This does not mean that an animal with primary consciousness cannot have long-term memory or act on it. Obviously, it can, but it cannot, in general, be aware of that memory or plan an extended future for itself based on that memory.

Only in ourselves—and to some extent in apes—does a higher-order consciousness emerge. Higher-order consciousness arises *from* primary consciousness—it supplements it, it does not replace it. It is dependent on the evolutionary development of language, together with the evolution of symbols, of cultural exchange; and with all this brings an unprecedented power of detachment, generation, and reflection, so that finally self-consciousness is achieved, the consciousness of being a self in the world, with human experience and imagination to call upon.

Higher-order consciousness releases us from the thrall of the here and now, allowing us to reflect, to introspect, to draw upon culture and history, and to achieve by these means a new order of development and mind. No other theorist I know of has even attempted a biological understanding of this step. To become conscious of being conscious, Edelman stresses, systems of memory must be related to representation of a self. This is not possible unless the contents, the 'scenes', of primary consciousness are subjected to a further process and are themselves recategorized.

Though language, in Edelman's view, is not crucial for the development of higher-order consciousness (there is some evidence of higher-order consciousness and self-consciousness in apes), it immensely facilitates and expands this by making possible previously unattainable conceptual and symbolic powers. Thus two steps, two re-entrant processes, are envisaged here: first the linking of primary (or 'value-category') memory with current perception—a perceptual 'bootstrapping', that creates primary consciousness; second, a linking between symbolic memory and conceptual centres—the 'semantic bootstrapping' necessary for higher consciousness. The effects of this are momentous: 'The acquisition of a new kind of memory', Edelman writes, '. . . leads to a conceptual explosion. As a result, concepts of the self, the past, and the future can be connected to primary consciousness. "Consciousness of consciousness" becomes possible.'[14]

At this point Edelman makes explicit what is implicit throughout his work—the interaction of 'neural Darwinism' with classical Darwinism. What occurs 'explosively' in individual development must have been equally critical in evolutionary development. Thus 'at some transcendent moment in evolution', Edelman writes, there emerged 'a variant with a re-entrant circuit linking value category memory' to current perception. 'At this moment,' Edelman continues, 'memory became the substrate and servant of consciousness.' And then, at another transcendent moment, by another, higher turn of re-entry, higher-order consciousness arose.

There is indeed much palaeontological evidence that higher-order consciousness developed in an astonishingly short space of time—some tens (perhaps hundreds) of thousands of years, not the many millions usually needed for evolutionary change. The speed of this development has always been a most formidable challenge for evolutionary theorists; Darwin himself could offer no detailed account of it, and Wallace was driven back to thoughts of a grand design.

The principles underlying brain development and the mechanisms outlined in the Theory of Neuronal Group Selection can, Edelman argues, account for this rapid emergence, since they allow for enormous changes in brain size over the relatively short evolutionary period in which *Homo sapiens* emerged. according to topobiology, relatively large changes in the structure of the brain can occur through changes in the genes that regulate the brain's morphology, changes that can come about as the result of relatively few mutations. And the premises of the theory of neuronal group selection allow for the rapid incorporation into existing brain structures of new and enlarged neuronal maps with a variety of functions.

3

New theories arise from a crisis in scientific understanding, an acute incompatibility between observations and existing theories. There are many such crises in neuroscience today. Edelman, with his background in morphology and development, speaks of the 'structural' crisis, the now well-established fact that there is no precise wiring in the brain, that there are vast numbers of unidentifiable inputs to each cell, and that such a jungle of connections is incompatible with any simple computational theory. But, at the other extreme, he is also moved, as William James was, by the apparently seamless quality of experience and consciousness—the unitary appearance of the world to a perceiver, despite (as we have seen in regard to vision) the multitude of discrete and parallel systems for perceiving it; and the fact that some integrating or unifying or 'binding' must occur, which is totally inexplicable by any existing theory.

Since the theory of neuronal group selection was first formulated, important new evidence has emerged suggesting how widely separated groups of neurones in the visual cortex can become synchronized and respond in unison when an animal is faced with a new perceptual task—a finding directly suggestive of re-entrant signalling.[15] There is also much evidence of a more clinical sort, which one feels may be illuminated, and perhaps explained, by the theory of neuronal group selection.

I often encounter situations in day-to-day neurological practice which completely defeat classical neurological explanations, which cry out for explanations of a radically different kind, and which are clarified by Edelman's theory.[16] Thus if a spinal anaesthetic is given to a patient—as used to be done frequently to women in childbirth—there is not just a feeling of numbness below the waist. There is, rather, the sense that one terminates at the umbilicus, that one's corporeal self has no extension below this, and that what lies below is not-self, not-flesh, not-real, not-anything. The anaesthetized lower half has a bewildering nonentity, completely lacks meaning and personal reference. The baffled mind is unable to categorize it, to relate it in any way to the self. One knows that sooner or later the anaesthetic will wear off, yet it is impossible to imagine the missing parts in a positive way. There is an absolute gap in primary consciousness which higher-order consciousness can report, but cannot correct. Since human beings never experience their own primary consciousness 'raw', but only as it has been transformed and enlarged by higher-order consciousness, the terms in which we experience such a gap become conceptual: thus 'alien' entails an explicit concept of 'self' and 'nothingness', and explicit concept of 'being'; such experiences allow a sort of clinical ontology.[17]

This indeed is a situation I know well from personal no less than clinical experience, for it is what I experienced myself after a nerve injury to one leg, when for a period of two weeks, while the leg lay immobile and senseless, I found it alien, not mine, not me, not real. I was astonished when this happened, and unassisted by my neurological knowledge: the situation was clearly neurological, but classical neurology has nothing to say about the relation of sensation to knowledge and to 'self'; about how, normally, the body is 'owned'; and how, if the flow of neural information is impaired, it may be lost to consciousness, and 'disowned'—for it does not see consciousness as a process.[18]

Such body-image and body-ego disturbances can be fully understood, in Edelman's thinking, as breakdowns in local mapping, consequent upon nerve damage or disuse. It has been confirmed, further, in animal experiments, that the mapping of body-image is not something fixed, but plastic and dynamic, and dependent upon a continual inflow of experience and use; and that if there is continuing interference with, say, one's perception of a limb or its use, there is not only a rapid loss of its cerebral map, but a rapid remapping of the rest of the body which then excludes the limb itself.[19]

Stranger still are the situations which arise when the cerebral basis of body-image is affected, especially if the right hemisphere of the brain is badly damaged in its sensory areas. At such times patients may show an 'anosognosia', an unawareness that anything is the matter, even though the left side of the body may be senseless, and perhaps paralysed, too. Or they may show a strange levity, insisting that their own left sides belong to someone else. Such patients may behave (as the eminent neurologist, M.-M. Mesulam, has written) '. . . as if one half of the universe had abruptly ceased to exist . . . as if nothing were actually happening [there] . . . as if nothing of importance could be expected to occur there'. Such patients live in a hemi-space, a bisected world, but for them, subjectively, their space and world is entire. Anosognosia is unintelligible (and was for years misinterpreted as a bizarre neurotic symptom) unless we see it (in Edelman's term) as 'a disease of consciousness', a total breakdown of high-level re-entrant signalling and mapping in one hemisphere—the right hemisphere, which, Edelman suggests, may have only primary but no higher-order consciousness—and a radical reorganization of consciousness in consequence.

Less dramatic than these complete disappearances of self or parts of the self from consciousness, but still remarkable in the extreme, are situations in which, following a neurological lesion, a dissociation occurs between perception and consciousness, or memory and consciousness, cases in which there remain only 'implicit' perception or knowledge or memory. Thus my amnesiac patient Jimmie ('the Lost Mariner') had no explicit memory of

Kennedy's assassination, and would indeed say, 'No president in this century has been assassinated, that I know of'. But if asked, 'Hypothetically, then, if a presidential assassination had somehow occurred without your knowledge, where might you guess it occurred: New York, Chicago, Dallas, New Orleans, or San Francisco?' he would invariably 'guess' correctly, Dallas.

Similarly, patients with visual agnosias, like Dr P. ('the Man who Mistook his Wife for a Hat'), while not consciously able to recognize anyone, often 'guess' the identity of people's faces correctly. And patients with total cortical blindness, from massive bilateral damage to the primary visual areas of the brain, while asserting that they can see nothing, may also mysteriously 'guess' correctly what lies before them: so-called 'blindsight'. In all these cases, then, we find that perception, and perceptual categorization of the kind described by Edelman, has been preserved, but has been divorced from consciousness.

In such cases it appears to be only the final process, in which the re-entrant loops combine memory with current perceptual categorization, that breaks down. Their understanding, so elusive hitherto, seems to come closer with Edelman's 're-entrant' model of consciousness.

Dissatisfaction with the classical theories is not confined to clinical neurologists; it is also to be found among theorists of child development, among cognitive and experimental psychologists, among linguists, and among psychoanalysts. All find themselves in need of new models. This was abundantly clear at the conference on 'Selectionism and the Brain'. Particularly suggestive, for me, has been the work of Esther Thelen and her colleagues at the University of Indiana in Bloomington, who have for some years been making a minute analysis of the development of motor skills— walking, reaching for objects—in infants. 'For the developmental theorist', Thelen writes, 'individual differences pose an enormous challenge . . . Developmental theory has not met this challenge with much success'. And this is, in part, because individual differences are seen as extraneous, whereas Thelen argues that it is precisely such differences, the huge variation between individuals, that allow the evolution of unique motor patterns.

Thelen finds that the development of such skills, as Edelman's theory would suggest, follows no single programmed or prescribed pattern. Indeed there is great variability among infants at first, with many patterns of reaching for objects; but there then occurs, over the course of several months, a competition among these patterns, a discovery or selection of workable patterns, or workable motor solutions. These solutions, though roughly similar (for there are a limited number of ways in which an infant can reach), are always different and individual, adapted to the particular dynamics of each child, and they emerge by degrees, through exploration and trial. Each child, Thelen has shown, explores a rich range of possible ways to reach for an object and selects its own path, without the benefit of any blueprint or program. The child is forced to be original, to create its own solutions. Such an adventurous course carries its own risks—the child may evolve a *bad* motor solution—but sooner or later such bad solutions tend to destabilize, break down, and make way for further exploration, and better solutions.[20]

When Thelen tries to envisage the neural basis of such learning, she uses terms very similar to Edelman's; she sees a 'population' of movements being selected or 'pruned' by experience. She writes of infants 'remapping' the neuronal groups that are correlated with their movements, and 'selectively strengthening particular neuronal groups'. She has, of course, no direct evidence for this, and such evidence cannot be obtained until we have a way of visualizing vast numbers of neuronal groups simultaneously in a conscious subject, and following their interactions for months on end. No such visualization is possible at the present time, but it will perhaps become possible by the end of the decade. Meanwhile, the close correspondence between Thelen's observations and the kind of behaviour that would be expected from Edelman's theory is striking.

If Esther Thelen is concerned with direct observation of the development of motor skills in the infant, Arnold Modell of Harvard has been concerned with psychoanalytical interpretations of early behaviour; he too feels, like Thelen, that a crisis has developed, that it might also be resolved by the theory of neuronal group selection: indeed, the title of his paper is 'Neural Darwinism and a conceptual crisis in psychoanalysis'. The particular crisis he spoke of was connected with Freud's concept of *Nachtraglichkeit*, the retranscription of memories that had become pathologically fixed, fossilized, but were now opened to consciousness, to new contexts and reconstructions, as a crucial part of the therapeutic process of liberating the patient from the past, allowing him to experience and move freely once again.

This process cannot be understood in terms of the classical concept of memory, in which a fixed record or trace or representation is stored in the brain—an entirely static or mechanical concept—but requires a concept of memory as active and 'inventive'.[21] That memory is essentially constructive (as Coleridge insisted, nearly two centuries ago)[22] was shown experimentally by the great Cambridge psychologist Frederic Bartlett. 'Remembering', he wrote, 'is not the re-excitation of innumerable fixed, lifeless and fragmentary traces. It is an imaginative reconstruction, or construction, built out of the relation of our attitude toward a whole mass of organized past reactions or experience.'

It was just such an imaginative, context-dependent construction or reconstruction that Freud meant by *Nachtraglichkeit*—but this, Modell emphasizes, could not be given any biological basis until Edelman's notion of memory as recategorization. Beyond this, Modell as an analyst is concerned with the question of 'self-creation'—how the self is created, its development and growth through finding, or making, personal meanings. Such a form of inner growth, so different from 'learning' in the usual sense, he feels, may also find its neural basis in the formation of ever-richer but always self-referential maps in the brain, and their incessant integration through re-entrant signalling, as Edelman has described it.[23]

Others too—cognitive psychologists and linguists—have become intensely interested in Edelman's ideas, in particular by the implication of the extended theory of neuronal group selection which suggests that the exploring child, the exploring organism, seeks (or imposes) meaning at all times, that its mappings are mappings of meaning, that its world and (if higher consciousness is present) its symbolic systems are *constructed* of, 'meanings'. When Jerome Bruner and others launched the 'cognitive revolution' in the mid-1950s, this was in part a reaction to behaviourism and other 'isms' which denied the existence and structure of the mind. The cognitive revolution was designed 'to replace the mind in nature', to see the seeking of meaning as central to the organism. In his book, *Acts of meaning*, Bruner describes how this original impetus was subverted, and replaced by notions of computation, information processing, etc., and by the computational (and Chomskian) notion that the syntax of a language could be separated from its semantics.[24]

But, as Edelman writes, it is increasingly clear, from studying the natural acquisition of language in the child, and, equally, from the persistent failure of computers to 'understand' language, its rich ambiguity and polysemy, that syntax cannot be separated from semantics. It is precisely through the medium of 'meanings' that natural language and natural intelligence are built up. From Boole, with his 'Laws of Thought' in the 1850s, to the pioneers of artificial intelligence at the present day, there has been a persistent notion that one may have an intelligence or a language based on pure logic, without anything so messy as 'meaning' being involved. That this is not the case, and cannot be the case, may now find a biological grounding in the theory of neuronal group selection.

None of this, however, can yet be proved: we have no way of seeing neuronal groups or maps or their interactions; no way of listening to the re-entrant orchestra of the brain. Our capacity to analyse the living brain is still far too crude. Partly for this reason researchers in neuroscience have felt it necessary to simulate the brain, and the power of computers and supercomputers makes this more and more possible. One can endow simulated neurones with physiologically realistic properties, and allow them to interact in physiologically realistic ways.

Edelman and his colleagues at the Neurosciences Institute have been deeply interested in such 'synthetic neural modelling', and have devised a series of 'synthetic animals' or artefacts designed to test the theory of neuronal group selection. Although these 'creatures'—which have been named Darwin I, II, III, and IV—make use of supercomputers, their behaviour (if one may use the word) is not programmed, not robotic, but (in Edelman's word) 'noetic'. They incorporate both a selectional system and a primitive set of 'values'—for example, that light is better than no light—which generally guide behaviour but do not determine it or make it predictable. Unpredictable variations are introduced in both the artefact and its environment so that it is forced to create its own categorizations.

Darwin IV or NOMAD, with its electronic eye and snout, has no 'goal', no 'agenda', but resides in a sort of pen, a world of varied simple objects (with different colours, shapes, textures, weights). True to its name, it wanders around like a curious infant, exploring these objects, reaching for them, classifying them, building with them, in a spontaneous, idiosyncratic way (the movement of the artefact is exceedingly slow, and one needs time-lapse photography to bring home its creatural quality). No two 'individuals' show identical behaviour; and the details of their reachings and learnings cannot be predicted, any more than Thelen can predict the development of particular movement styles in her infants. If their value circuits are cut, the artefacts show no learning, no 'motivation', no convergent behaviour at all, but wander around in an aimless way, like patients who have had their frontal lobes destroyed. Since the entire circuitry of these Darwins is known, and can be seen functioning in detail on the screen of a supercomputer, one can continuously monitor their inner workings, their internal mappings, their re-entrant signallings: one can see how they sample the environment, how the first, vague, tentative percepts emerge, and how, with hundreds of further samplings, they evolve and become recognizable, refined models of reality, following a process similar to that projected by Edelman's theory.[25]

Seeing the Darwins, especially Darwin IV, at work can induce a curious state of mind. Going to the zoo after my first sight of Darwin IV, I found myself looking at birds, antelopes, lions, with a new eye: were they, so to speak, nature's Darwins, somewhere up around Darwin XII in complexity? And the gorillas, with higher-order consciousness but no language—where would they stand? Darwin XIX? And we, writing about the gorillas, where would we stand? Darwin XXVII, perhaps? Edelman often wonders about the possibility of constructing a conscious artefact—he has no doubt of the possibility, but places it, mercifully, well on in the next century.

Neural Darwinism (or Neural Edelmanism, as Francis Crick has called it) coincides with our sense of 'flow', that feeling we have when we are functioning optimally, of a swift, effortless, complex, ever-changing, but integrated and orchestrated stream of consciousness;[26] it coincides with the sense that this consciousness is ours, and that all we experience and do and say is, implicitly, a form of self-expression, and that we are destined, whether we wish it or not, to a life of particularity and self-development; it coincides, finally, with our sense that life is a journey—unpredictable, full of risk and uncertainty, but, equally, full of novelty and adventure, and characterized (if not sabotaged by external constraints or pathology) by constant advance, an ever-deeper exploration and understanding of the world.

Edelman's theory proposes a way of grounding all this in known facts about the nervous system and testable hypotheses about its operations. Any theory, even a wrong theory, is better than no theory; and this theory should at least stimulate a storm of experiment and discussion—for it is the first truly global theory of mind and consciousness, the first biological theory of individuality and autonomy.

NOTES AND REFERENCES

1. The heady atmosphere of these days is vividly captured in *The Cybernetics Group* by Steve J. Helms (MIT Press, 1991), and many of the McCulloch papers were later collected in *Embodiments of mind* (MIT Press, 1965).

2. See Marvin Minsky's 1967 book, *Perceptrons*, in this regard.

3. See Francis Crick (1989). The recent excitement about neural networks. *Nature*, **337**, 129–32.

4. Though, increasingly, the term 'adaptive systems' is used for automata or complex systems that can modify themselves in relation to assigned 'tasks' or 'challenges', and in a certain (but unbiological) sense be said to 'adapt' or 'learn'. Such adaptive systems, of ever-increasing sophistication, are proving of major importance in many realms, and have been an especial concern of the researchers at the Santa Fe Institute.

5. Lashley expressed this in a famous paper, 'In search of the engram', published shortly before his death; London: *Symposia of the Society for Experimental Biology*, **4** (1950).

6. Jeffrey Gray's article is to be found in *Nature*, Vol. 358 (July 1992), p. 277, and my own reply to it in *Nature*, Vol. 358 (August 1992), p. 618.

7. Edelman first presented this in a relatively brief essay written in 1978 (*The mindful brain*, MIT Press). This essay was written, Edelman has said, in a single sitting, during a thirteen-hour wait for a plane at Milan airport, and it is fascinating to see in it the germ of all his future thought: one gets an intense sense of the evolution occurring in him. Between 1987 and 1990 Edelman published his monumental and sometimes impenetrable trilogy *Neural Darwinism* (1987), *Topobiology* (1988), and *The remembered present: a biological theory of consciousness* (1989), which presented the theory, and a vast range of relevant observations in a much more elaborate and rigorous form. He presents the theory more informally, but within a

richer historical and philosphical discussion, in his most recent book, *Bright air, brilliant fire* (Basic Books, New York, 1992).

8. In 'To see and not see' (*The New Yorker*, 10 May 1993), I describe not the personal world of an infant (of which we can never hope to get any report), but of a newly sighted adult. This man found himself deluged with raw visual sensations, and was at first wholly unable to recognize or discriminate people, objects, or even simple geometrical shapes. Nothing has shown me more clearly the crucial role of experience in the development of our perceptual capacities.

9. And yet, clearly, much more complex capacities and dispositions may be 'built-in' to the genotype of the organism, although their development (or lack of development) may depend on experience. This is so for our species-specific linguistic capacity, our varied intellectual capacities (musical capacity is one of the most innate and specific), and may subtle dispositions and behavioural traits. This may be especially striking in identical twins who have been separated at birth.

10. D. Stern. *The interpersonal world of the infant: a view from psychoanalysis and developmental psychology* (Basic Books, New York, 1985).

11. There may, however, be built-in mechanisms for certain generic recognitions, such as the ability, which we share with all primates, to recognize the category of 'snakes' even if we have never seen a snake before; or infants' ability to recognize the generic category of 'faces' long before they recognize particular ones. There is now evidence for 'face-detecting' cells in the cerebral cortex.

12. Confusingly, the very term 're-entrant' has occasionally been used in the past to denote such feedback loops. (It is used in this way by McCulloch in his early papers on automata.) Edelman gives the term 're-entry' a radically new meaning.

13. Thus if pigeons are presented with photographs of trees, or oak leaves, or fish, surrounded by extraneous features, they rapidly learn to 'home in' upon these, and to generalize, so that they can thereafter recognize any trees, or oak leaves, or fish straight away, however distracting or confusing the context may be. It is clear from these experiments that perception selects, or rather creates, 'defining' features (what counts as 'defining' may be different for each pigeon), and cognitive categories, without the use of language, or being 'told' what to do. Such category-creating behaviour (which Edelman calls 'noetic') is very different from the rigid, algorithmic procedures used by robots. (These experiments with pigeons are described in detail in Edelman's *Neural Darwinism*, pp. 247–51).

14. This explosion normally occurs in the third year of life, and is spread over several months as language is acquired. But if through special circumstances—deafness, incarceration (as with Kaspar Hauser), or lack of contact with other human beings (as with 'wild' children)—a child is denied contact with language at this age, and develops it only later, the development of higher-order consciousness may be truly explosive, and may occur in a matter of hours or days, with the sudden, belated rushing-in of language. (See *Seeing voices*, pp. 45–58 (O. Sacks, Picador, 1990).)

15. I discussed this work in an earlier article, 'Neurology and the Soul', *The New York Review of Books*, 22 November 1990.

16. Some of these situations are discussed by Israel Rosenfield in his new book *The strange, familiar and forgotten*, where he speaks of 'the bankruptcy of classical neurology'.

17. Animals without higher-order consciousness or self-consciousness show no sign of recognizing self-absence or nothingness, and if the hindquarters of such an animal are anaesthetized, the animal will simply ignore them, and go about its business, without showing any signs of bewilderment or that it perceives anything amiss. This is well described by the veterinarian James Herriott, in regard to a cow given a spinal anaesthetic for obstructed calving; it became indifferent to the whole procedure as soon as the anaesthetic took effect, and returned to munching its hay.

18. A full discussion of such body-image or body-ego disturbances in relation to TNGS I can be found in a new afterword to the UK edition of my book *A leg to stand on* (Picador, 1992).

19. Fundamental work showing the plasticity of the cerebral cortex, and the remarkable degree to which it can reorganize itself after injuries, amputations, strokes, etc., has been done by Michael Merzenich and his colleagues at the University of California in San Francisco. See (for example): 'Cortical representational plasticity', by M.M. Merzenich, G. Recanzone, W.M. Jenkins, T.T. Allard, and R.J. Nudo in *Neurobiology of the neocortex*, edited by P. Rakic and W. Singer (John Wiley, New York, 1988), pp. 41–67.

20. See Esther Thelen, 'Dynamical systems and the generation of individual differences', in *Individual differences in infancy: reliability, stability, and prediction*, edited by J. Colombo and J.W. Fagen (Erlbaum, Hillsdale, New Jersey, 1990). Similar considerations arise with regard to recovery and rehabilitation after strokes and other injuries. There are no rules, there is no prescribed path of recovery; every patient must discover, or create, his own motor and perceptual patterns, his own solutions to the challenges that face him; and it is the function of a sensitive therapist to help him in this.

21. See Israel Rosenfield, *The invention of memory: a new view of the brain* (Basic Books, New York, 1991).

22. No one liked this distinction, or articulated it more eloquently, than Coleridge, who saw himself as 'laying the foundation Stones of Constructive or Dynamic Philosophy in opposition to the merely mechanic' (in his letter to Brabant, 21 July 1815). This, for him, meant transforming the concepts of perception, memory, knowledge, and imagination, or rather, seeing that these were of two sorts— which at various times he calls passive and active, dead and alive, false and true. In one of his notebooks (II: 1509), he speaks of 'mock' knowledge ('having no roots . . . no buds or twigs . . . but a dry stick of licorish') as opposed to true knowledge, which is rooted and grows and lives within one. And in his *Biographia literaria*, in a famous passage which might have served as an inspiration to Bartlett, he contrasts the constructive (Imagination) with the merely mechanic (Fancy): 'The Imagination . . . dissolves, diffuses, dissipates, in order to recreate . . . it struggles to idealize and to unify. It is essentially *vital*, even as all objects (as objects) are essentially fixed and dead. Fancy, on the contrary, has no other counters to play with, but fixities and definites . . . [it] must receive all its materials ready made from the law of association.'

23. Modell's ideas have been set out in full in *Other times, other realities* (Harvard University Press, 1990), and in *The private self* (Harvard University Press, 1993).

24. Jerome Bruner, *Acts of meaning* (Harvard University Press, 1990).

25. Normally one is not aware of the brain's almost automatic generation of 'perceptual hypotheses' (in Richard Gregory's term) and their refinement through a process of repeated samplings and testing. But under certain circumstances, as in recovery after acute nerve injury, one may become vividly aware of these normally unconscious (and sometimes exceedingly rapid) operations. I give a personal example of this in *A leg to stand on* (see note 18). One is much more aware of such hypothesizing when sensory information is scanty or ambiguous—as, for example, when driving in unfamiliar terrain at night.

26. See Mihaly Csikszentmihalyi, *Flow: the psychology of optimal experience* (Harper Collins, New York, 1990).

CHAPTER EIGHT

The limitless power of science

P. W. ATKINS

I shall explore what I see as the scope of science in explaining the structure and events in the world and ask whether it is limitless. It is therefore inevitable that I speak of religion too, and of its claims that it is an essential component of our understanding of things.

I shall begin with religion's place in the world of explanation. The inclination of the religious seems to be to hope to accommodate the discoveries of science, and in doing so to find a richer comprehension of the world. No religious person, the religious claim, should fear the apparent tensions of science and religion, for their conflict is superficial. In due course it will be seen, it is claimed, that science and religion are partners in a joint activity: the two streams of enquiry and revelation will merge and be seen to be mutually enriching, not mutually annihilating. A well-rounded view of the world will be obtained, the religious claim, only if we listen to science, and its message about its zone of competence, as closely as we listen to the Bible; for science is said to elaborate the Bible. Religion is the antidote to reductionism, for it illuminates the whole rather than the fragments of comprehension.

That, I believe, summarizes in a general way the stance of the scientifically articulate religious believer. It is attractive because it has an air of humility laced with generosity, and it appeals to those who take satisfaction from arguments that conclude in the concord of harmonious compromise. An alternative point of view, however, is that religion has had its day and that science, and the tracing of phenomena to its atomic roots that epitomizes reductionism, should be regarded as supreme. It may be that the religious (and the faint-hearted among agnostic or disinterested and busy scientists) are seeking a final compromise with science. That compromise, which is urged on us when we are encouraged to divide up domains of enquiry, is the last vain attempt to accommodate the conqueror before religion loses all claim to the respect of minds and pretence of truth. I consider that the survival of religion and the antireductionism that it represents survives merely because it is so deeply ingrained in our cultural attitudes, and its survival is independent of its intrinsic truth. The stifling grip of religion on Man's mind stems partly from its early start, when, as our ancestors dropped from the trees they first sought explanations and solace; it also stems partly from religion's control (for both benevolent and malevolent purposes) of the behaviour of individuals and societies, and it stems partly from its capture of the literature and the arts, which has given it a powerful imagery. Someone with a fresh mind, one not conditioned by upbringing and environment, would doubtless look at science and the powerful reductionism that it inspires as overwhelmingly the better mode of understanding the world, and would doubtless scorn religion as sentimental wishful thinking. Would not that same uncluttered mind also see the attempts to reconcile science and religion by disparaging the reduction of the complex to the simple as attempts guided by muddle-headed sentiment and intellectually dishonest emotion?

But, the antireductionists will cry, we need to listen to the spirit and that science cannot replace. Now, though, you speak of the illumination provided by poets, and although poets may aspire to understanding, their talents are more akin to entertaining self-deception. They may be able to emphasize delights in the world, but they are deluded if they and their audience believe that their identification of the delights and their use of poignant language are enough for comprehension. Philosophers, too, I am afraid, have contributed to the understanding of the Universe little more than poets. They have raised questions, examined the frailties and imprecisions of human languages, and have worried a great deal about what may be a question, but they have not contributed much that is novel until after novelty has been discovered by scientists.

Theologians, incidentally, have contributed nothing. They have invented a world and language of their own, like some mathematicians, but unlike many mathematicians have sought to impose its percepts and precepts on this world. In so doing they have contaminated truth, and wasted the time of those who wish to understand this world. Scientists have had and are continuing to have to scrape away the detritus of religious obfuscation before they can begin their own elucidation.

Scientists, with their implicit trust in reductionism, are privileged to be at the summit of knowledge, and to see further into truth than any of their contemporaries. They are busy in the public domain, where truth can be tested by shared experience, where truth supervenes international boundaries and cultures. Scientists liberate truth from prejudice, and through their work lend wings to society's aspirations. While poetry titillates and theology obfuscates, science liberates.

The grave responsibility of scientists is to use their voices to blow back the fog that shrouds the minds of those who have not yet seen. Scientists are successfully treading the path of reductionism. They are exposing the simple essentials of the world, seeing its mechanism, seeing that they can comprehend its actions, and seeing that they can understand its origin and elucidate the problems that have puzzled people and given priests their power.

Scientists have a duty to reveal to the public their insight into the world's mechanism. They are the beacons of rationality, lighting the trail for those who wish to use that most powerful and precious of devices, the human brain. They light the beacons for those who wish to escape the prejudice of those who, through irrational belief and faith in the ultimately unknowable, live lives blighted by others.

Theism (and the implicit rejection of reductionism) is a system of knowledge based on ignorance, and that twin of ignorance, fear. It would certainly be too much to expect a theologian (or indeed a scientist) to admit that his lifetime's work had been based on a false foundation. It is even less likely that anyone religious, unless they were exceptionally self-honest and intellectually sinewy, would admit that the whole history of their church was based on a clever, but understandable, self-delusion (and in some cases, I suspect, on a straightfor-ward conscious lie). I consider that religion is a delusion propagated by a combination of ignorance, art, and fear, fanned into longevity and ubiquity by the power it gave to those in command.

Religion emerged from magic, and has never completely discarded its origins. It is a sophistication and institutionalization of magic. It is an elaboration of our bewildered ancestor's belief that there are forces to be cajoled and appeased. Its present power stems from the grip it exerts on minds with its amelioration of the prospect of death, and from its ability to provide sustenance for those exposed to the hardships of life. In its quest for understanding and its intention to help, religion is good. But at root, being founded on ignorance, it is baseless. Religion is utterly unable to digest the paradigm of reductionism, for it rejects the possibility of overall understanding that will come from the reduction of the complexity of the world to its simple substructure and the awesomely difficult tracking of the interrelations of that simplicity back into the complexity of perceived events.

Let me contrast the styles of evidence for theistic and scientific explanations. Suppose you have a blank, unprejudiced mind and are presented with the evidence for religion. You are shown some very beautiful words, and some equally beautiful buildings. Many of the intentions you are shown are highly laudable. At the centre of the evidence you are shown a great and ornate cabinet, a cabinet encrusted with love and time and labelled Faith. You are told that everything outside the cabinet is not really central, that the most compelling evidence, the irrefutable evidence, the evidence to out-jury any jury lies in that cabinet of Faith. However, you are also told, amid assurances that the evidence is indeed inside, and very, very compelling, that to inspect the interior of the glorious cabinet it is necessary to enter through the door marked Death. What adult, rational intellect would accept such hypocrisy? Only, I claim, the unthinking, the intellectually dishonest, and the people who cannot come to terms with the prospect of their own annihilation.

Science, the system of belief founded securely on publicly shared reproducible knowledge, emerged from religion. As science discarded its chrysalis to become its present butterfly, it took over the heath. There is no reason to suppose that science cannot deal with every aspect of existence. Only the religious—among whom I include not merely the prejudiced but also the underinformed—hope that there is a dark corner of the physical Universe, or of the universe of experience, that science can never hope to illuminate. But science has never encountered a barrier, and the only grounds for supposing that reductionism will fail are pessimism in the minds of scientists and fear in the minds of the religious. The frightened seek to erect false barriers, and vainly hope to preserve their gods from annihilation by defining different domains of competence for science and religion, and by pretending that science is incompetent when it brings its razor to bear on belief.

Religion closes off the central questions of existence by attempting to dissuade us from further enquiry by asserting that we cannot ever hope to comprehend. We are, religion asserts, simply too puny. Through fear of being shown to be vacuous, religion denies the awesome power of human comprehension. It seeks to thwart, by encouraging awe in things unseen, the disclosure of the emptiness of faith. Religion, in contrast to science, deploys the repugnant view that the world is too big for our understanding. Science, in contrast to religion, opens up the great questions of being to rational discussion, to discussion with the prospect of resolution and elucidation. Science, above all, respects the power of the human intellect. Science is the apotheosis of the intellect and the consummation of the Renaissance. Science respects more deeply the potential of humanity than religion ever can.

Simple ideas distinguish science so gloriously from religion, and science's self-sufficiency, its ability to deal with any problem without needing to import some external mysterious cause. Science identifies simple concepts that effloresce brilliantly and publicly into testable conclusions. That is high reductionism. Religion expresses indefinite, obscure ideas that are sometimes a practical (but not always a beneficial) foundation for ethics and morality, but beyond being a guide for human behaviour (for good or for ill) and a salve for the oppressed (and sometimes a weapon of oppression) have no success, utterly none, in the accounting for, let alone predicting, the phenomena of the world. That is the legacy of no reductionism.

There are, to my mind, two great aspects of science in which humanity should take delight. One is the connection science exposes between the disparate. Science, by pursuing the reduction of observation to underlying simplicities, exposes the unity of the world. We see through reductionist science that knowing one thing explains many. As we untangle the composition of Nature we see that it conjures differences by the simple modification of a chemical bond or the deployment of a different atom. It is so important to understand that science identifies extremely simple concepts of exceptional richness and shows how they bind together the apparently disparate into a web. It is the task of science to identify the seeds of explanation, and to plant them like acorns into minds. There they will grow, and people and we among them will come to see that the explanations of the perceived world are as an oak to its acorn.

The second feature of science is that it shows that the world is simple. Even many scientists do not appreciate that they are hewers of simplicity from complexity. They are often more deluded than those they aim to tell. Scientists are often overawed by the complexity of detecting simplicity. They look at the latest fundamental particle experiment, see that it involves a thousand kilograms of apparatus and a discernible percentage of a gross national product, and become thunderstruck. They see the complexity of the apparatus and the intensity of the effort needed to construct and operate it, and confuse that with the simplicity that the experiment, if successful, will expose. Some scientists are so awestruck that they even turn to religion! Others keep a cool head, and marvel not at an implied design but at the richness of simplicity.

Most scientists glance at a mathematical formula and see awesome complexity, and marvel at the brain that first derived it. Like priests, some then seek to retain the mysteries. The awe is well aimed, but nevertheless not quite the correct response. There should be awe reserved for the original discoverer, for few have the power to discover new continents, or even islands, of knowledge. There should be delight, not awe, for the recomfirmation that the human brain is such a brilliant instrument that it can make light of darkness. But most important of all, there should be realization that a connection and a simplicity have been exposed. The connection is the formula, which bundles several knowns together, and shows that they account for another known. The simplicity is the reduction of the concepts that the new relation implies, although this is often interpreted as a complexity.

I can give you an idea of what it is I mean by talking a little about one of the greatest insights of science, the Second Law of thermodynamics. As far as we are concerned, all this law states is that all events are accompanied by an increase in disorder. The world is slipping into chaos. Some are puzzled by that remark for they cannot see that people can emerge. Those benighted dark-bound over-prejudiced commentators the creationists think that the Second Law is a sign that Man must have been created, for how can ordered Man arise spontaneously from disordered slime? A scientists says 'good question', and then proceeds to answer it by showing that collapse into disorder may be constructive. Events do not occur in isolation, and if an event here creates much disorder, and is connected to a process there that leads to the elimination of less disorder, then the joint process may occur spontaneously because it leads to net disorder. Those who look at the flower opening, or the fish evolving, may see only the constructive process, and be awed. Those others of us who trace the connections and see not only the flower opening but also the decay elsewhere, those of us who see the entire process, see overall collapse, and stand in less awe. Our awe is now directed at the richness of the interconnection and the ramifications of chaos, not at the burgeoning, apparently purposeful, emergence of form.

Dispersal into disorder creates because it need not be uniformly smooth. A flood of chaos there may result in a surge of order here. The purposeless increase in disorder of the world is not a smoothly descending river of energy, but a choppy rapid, that may throw up a structured foam and an elaborate wave as it plunges down. That order may take the form of a protein formed by an enzyme driven ultimately by the energy of the Sun, or the construction of a strip of DNA. It may power the jaws of a cheetah and the emergence on its coat of the stripes of a zebra. Thus the Second Law may erupt into evolution, and stronger cheetahs and better camouflaged zebras may emerge, transitorily, as the universe globally spreads in disorder. Thus the 'Creation'—everything—emerges as chaos spreads.

A gross contamination of the reductionist ethic is the concept of purpose. Science has no need of purpose. All events at the molecular level that lies beneath all our actions, activities, and reflections are purposeless, and are accounted for by the collapse of energy and matter into ever-increasing disorder. The richness of the events of the world stems from the interconnection of the collapse into the network that, here and there, as at you and me, can throw up a local abatement of chaos that we interpret as an action or a deed. Behaviour is also ultimately collapse into disorder, even though that behaviour may on occasion be exquisite.

Natural selection, like the interpretation of the Second Law, is one of the most beautiful theories of science, for it is so economical yet powerful. That is, it is simple, yet admits as a consequence complexity. That is the hallmark of high science and the apotheosis of reductionism. From a simple precept (e.g. the unconscious tendency of genes to survive in the face of unconscious competition) the whole gamut of organisms in the creation can be seen to emerge. It is then entirely gratuitous (but science cannot demonstrate that it is incorrect) to suppose that evolution has been guided teleologically or that matter had some in-built tendency to aggregate complexity. Both speculations are entirely possible, just as other profligate speculations are also possible, such as the unsupported musings that this universe is constantly splitting into others that have no communication with us, or that there is a teapot in orbit around Mars.

Science can perform its elucidation without appealing to the shroud of obscurity of man-made artifice, including that supreme artifice the presumption of purpose. A block of hot metal will cool, not because its purpose is to cool, but because the spontaneous chaotic dispersal of its energy results in its cooling. A shoot emerges from a seed and grows into a plant, not because the seed's purpose is to grow but because the intricate network of reactions in its cells are gearboxes that drive its growth as the rest of the world sinks a little more into chaos. The lily is a flag hoisted by collapse into purposeless chaos. All the extraordinary, wonderful richness of the world can be expressed as growth from the dunghill of purposeless interconnected corruption. People, too, have emerged as the same dunghill has effloresced. One molecule capable of reproducing itself in its own image is all it needs to set the world on the progress that culminates in it being peopled with persons.

Let me speak now a little of the attitudes that science inspires and adopts in order to discover more about the world. Science, as I have said, favours simplicity. Science is the arch-descendant of Ockam. How dare those theologians so obscure truth by their gildings, their hangings, their sentiment, their wishful thinking, their personal fears, and their network of intrusion into personal liberty! They have no right to claim that 'God' is an extreme simplicity, and as cogent and potent an explanation of our origins as is necessary. A 'God' is the embodiment of complexity, the ultimate antisimplicity. Maintaining that God is an explanation (of anything, not merely Creation) is an abnegation of the precious human power of reasoning out comprehension.

When confronted with the analysis of any concept, however complex, the only intellectually honest attitude to adopt is one of exploring the extent to which an absolutely minimal approach will prove sufficient to account for the reliable evidence. There is no justification for a departure from this attitude when the complex concept is that of our own origin or cosmic purpose, however deeply emotive and close to our sensibilities the subject may be and however we may long for a comforting outcome.

In seeking to understand our origin and our purpose science examines whether an absolutely minimal approach is sufficient. Only if a minimal approach is explicitly demonstrated to be inadequate may there be some justification in indulging in the soft furnishings of additional hypotheses. Science explores, and is having success in showing that our individual existence, persistence, and role in this universe can be explained without the accretion of invented hypothetical additions. Science shows that all our attributes can be explained without the sugar coating of invented attributes that have been proposed by the underinformed or the wily and have been adhered to generally by the religious.

Science is in the midst of showing that the concept of existence can survive the absolute simplicity of stripped-down explanations and their ramifications. Is there any support for the existence of something beyond the absolutely sparse? Is there life beyond bones? Are the fat and tallow of religious and other forms of philosophical or psychological justification necessary and not merely desirable? Science is succeeding without them.

Science treads everywhere, and worms itself under the scabs that religion regards as protecting the special tender patches of human existence. The religious go to intellectual war to maintain that in some areas secularly inspired logic cannot tread. Yet reductionist science is omnicompetent. Science has never encountered a barrier that it has not surmounted or that we can at least reasonably suppose it has power to surmount and will in due course be equipped to do so. There is no explicitly demonstrated validity in the view that there are aspects of the universe closed to science. I can accept, given the success with which science has encroached on the territory once regarded as traditionally religion's, that many people hope its domain of competence will prove bounded, with things of the spirit on that side of the fence and things of the flesh on this. But until it is proved otherwise, there is no reason to suppose that science is incompetent when it brings its razor to bear on belief. Until the day that science is explicitly shown to be incompetent, we should acknowledge that its not-yet-stopped reductionist razor is slicing through the fabric of the heavens, and is leading us towards an extraordinary deep understanding of the composition, organization, and origin of the world.

Science is slowly equipping itself to deal with aesthetic and religious experiences, and will be able to account for the perception of oneself as a distinct but responding entity. It will do so, I do not doubt, by showing that these characteristically human capacities, which we lump together for convenience of discourse as 'human spirit' or 'soul', are no more than psychological states of the brain. Likewise, that other component of our existence, the wishful thinking extension of the idea of 'soul' to the expectation of eternal persistence, is already quite plainly explicable in terms of the deep-seated desire to avoid, and the inability to come to terms with, the prospect of one's own annihilation.

There is no reason why consciousness should not be admitted into the kingdom of scientific explanation. Consciousness, like growth and digestion, is open to scientific explanation. There is no evidence that the brain's workings require non-physical impositions. The principal activity of the brain, that of sustaining a sense of consciousness through a lifetime, is open to explanation rooted in its physical structure (as governed by the body's genes being inherited and finding expression through the purposeless workings of the second Law, essentially building on the purposeless collapse of sandwiches into chaos) and its chemical activity (also the manifestation of the Second Law, in the same way). We can never hope to give a formula for the brain (whatever that might mean); but that is not the reductionist programme. Our programme is to comprehend its structure and mechanism to the point where we can build a reasonably accurate simulacrum. Such a brain might well have doubts about its role, an impression of purpose (in which it will be right), and fear that a switch can too lightly be thrown.

Once consciousness is admitted to the kingdom of science, then all our mental attributes are open to comprehension and not merely awe. The brain is subtle and capable it seems of infinite understanding of its self, its origin, its cosmic origin, the origin of the cosmos, and of that cosmos's immediate, intermediate, and long-term featureless future. However, like other complex instruments, the brain can make errors, especially when the opinions it sometimes expresses are pressed on it, and especially when they are widely supported by others. Thus the brain can hallucinate, and it can become perplexed, and armchair brains can avoid unnecessary exercise by adopting easy explanations. Some of these failures lead to poetic expression, others to religious fervour, and some to undisguised madness. All of them, though, are an abnegation of the brain's true power of understanding, which is to be found only in scientific explanation.

My scientific world-view is bleak in terms of its origins, its motivations, and its future. Yet, unless it can be explicitly demonstrated otherwise, it should be the sparse working hypothesis to account for our existence. If everything in the world can be accommodated in this bony view, then there is no justification to impose on our understanding the hypothetical extraneous. I challenge anyone who seeks, hopes, and believes in a seductively richer opinion and a rosier destiny, to travel to the bedrock of existence, and to build their beliefs on it only as their shelter is shown to be essential. I maintain that all softenings of my absolutely barren view of the foundations of this wonderful, extraordinary, and delightful world are sentimental wishful thinking.

Religion is argument by sentiment. Most thinking people would probably agree that the physical laws are best discovered by experiment, not sentiment. However, there will be those who see 'sentiment' as an antenna that responds to the side of nature left in shadow by science's emotionless glare, and sees what science cannot see. Let me stress again that I do not consider that there is any corner of the real universe or the mental universe that is shielded from this glare, and that there is no aspect of understanding where sentiment is a better source of understanding than science. This is particularly so in our understanding of the origin of ourselves and the Universe, including without exception its matter, energy, opinions, and its morals. Least of all do I favour the special pleading of a sentiment that drives arguments forward in a passionate chase towards the avoidance of annihilation by methods other than scientific and, that subset of science, medicine. I long for immortality, but I know that my only hope of achieving it is through science and medicine, not through sentiment and its subsets, art and theology.

Science is exceptionally honest, and succeeds because it bares its breast to constant attack. Take the goal it has for its description of the creation. If reductionist science is to be proved omnicompetent it must achieve a complete description, one devoid of precursor and intervention. That is, it must account for everything in the world, not only atoms, aardvarks, electrons, and lecturers, but also the sense of human spirit, goodwill, religious enthusiasm, and beliefs in Gods and the powers of potions and prayers. Moreover, if it is to be honest, it must achieve all this by starting from something without a precursor; that is, it must start from nothing at all. The scientific account of cosmogenesis cannot stop when it has accounted for the universe springing from a seed the size of a Sun, nor when it has arrived at a seed the size of a pea. Nor can it stop at any smaller seed. A seed the size of a proton implies that that seed had to be manufactured, placed there by some cosmic pre-existing gardener. Science will be forced to admit defeat if it has to stop at a seed of any size. That is the severity of the criterion that science sets for itself. If we are to be honest, then we have to accept that science will be able to claim complete success only if it achieves what many might think impossible: accounting for the emergence of everything from absolutely nothing. Not almost nothing, not a subatomic dust-like speck, but absolutely nothing. Nothing at all. Not even empty space.

How different this is from the soft flabbiness of a nonscientific argument, which typically lacks any external criterion of success except popular acclaim or the resignation of unthinking acceptance. One typical adipose arch-antireductionist argument may be that the world and its creatures were created (by something called a God), and that was that. Now, that may be true, and I cannot prove that it is not. However, it is no more than a paraphrase of the statement that 'the universe exists'. Moreover, if you read into it an active role for the God, it is an exceptionally complex explanation, even though it sounds simple, for it implies that everything (or almost everything, even if the God supplied no more than electrons and quarks) had to be provided initially.

My personal belief for the future of cosmology is that we shall shortly have to start thinking about what happened on the other side of zero. I suspect that the only route to knowledge about the structure and properties of the Universe and the processes that accompanied its inception is to speculate about the events that preceded the Creation. I see that we shall need to build a model of what was before there was time in a place not in space, and to explore whether its consequence was creation. We shall, in a sense, need to model nothing, and to see if its consequences are this world. I don't regard that as impossible or ludicrous; I regard it as the next logical step for the development of the paradigms of science.

The attitude that I advocate is that the omnicompetence of science, and in particular the simplicity its reductionist insight reveals, should be accepted as a working hypothesis until, if ever, it is proved inadequate. I began by wondering whether science and religion could be reconciled and if they were complementary explorations of the cosmos. I have to conclude that they cannot be reconciled. A scientists' explanation is in terms of a purposeless, knowable, and understandable fully reduced simplicity. Religion, on the other hand, seeks to explain in terms of a purposeful, unknowable, and incomprehensible irreducible complexity. Science and religion cannot be reconciled, and humanity should begin to appreciate the power of its child, and to beat off all attempts at compromise. Religion has failed, and its failures should stand exposed. Science, with its currently successful pursuit of universal competence through the identification of the minimal, the supreme delight of the intellect, should be acknowledged king.

REFERENCES

Atkins, P.W. (1992). *Creation revisited*. W.H. Freeman/Penguin.
Atkins, P.W. (1994). *The second law*. Scientific American Library; reissued 1994.
Dawkins, R. (1976). *The selfish gene*. Oxford University Press.

CHAPTER NINE

Reductive megalomania

MARY MIDGLEY

THE REDUCTIVE TEMPER

I want to examine the reductive approach as a general attitude, rather than looking at the forms of actual reductions. These forms vary. There can be reduction of wholes to parts, of mind to body, or of different conceptual schemes to one unifying one. But the use of the special word 'reductive' points to something they are believed to have in common. This element does not seem to be only a formal one. The point is not just that these are all ways of simplifying the conceptual scene. It concerns intentions, and examining those intentions is not an irrelevant, psychoanalysing exercise, because formal reductions don't spring up on their own, like weeds in a garden. They are not value-free. They are always parts of some larger enterprise, some project for reshaping the whole intellectual landscape, and often our general attitude to life as well. Whenever we get seriously involved in the business of reduction, either as supporters or resisters, we are usually responding to these wider projects.

My first point is that we need to be aware of these projects and to discuss them explicitly, even when they lead us far outside our specialities. We cannot settle the vast, looming, vulgar background questions indirectly by fixing the small logical issues. You can't shift a muck-heap with a teaspoon. My second point—about reductivism in particular—is that the large background projects involved are never just destructive but always aim at something positive as well. In our time, reduction overwhelmingly presents itself as purely negative, a mere exercise in logical hygiene, something as obviously necessary as throwing out the rubbish. But this presupposes that we have already made sure what we want to throw away and what we want to make room for.

Parsimony is a respectable ideal, but it is not one that makes sense on its own. In thought as in life, false economy is possible. We cannot tell what we should save until we have decided what we want to buy with our savings. Of course there do exist straightforward misers, savers led by pure parsimony. And there are intellectual misers too, sceptics who pride themselves on being too clever to believe anything or anybody. But most of us are not like that; we think of saving as a means to an end.

Reasoning does not actually demand the most economical account conceivable. It demands the most economical one *that will give us the explanation we need*. To get this, we need to consider carefully which lines to pursue: how wide our explanation needs to be, how large our question is and what other questions are bound up with it. Mapping these surrounding questions is an essential function of a good explanation.

Reducers, then, are quite entitled to make savings and to spend what they have saved on building explanatory structures elsewhere in order to balance those that they destroy. Things go wrong only when they do not notice that they are doing this and accordingly fail to criticize their own constructions. They think their work is much easier than it actually is because they feel sure in advance what they want to destroy. They often do not notice how much they are adding.

AUSTERITY?

Confidence of this kind can make for a strange complacency. In his preparatory notes for the conference, John Cornwell wrote of 'reductionism's austere outlook'. And pride in that austerity is obviously very widely felt by reducers. But there is no reason to think of reduction as necessarily austere. Intellectual puritans, like other kinds of puritan, are usually after a pay-off, an imaginative indulgence to compensate for their surface austerities. This pay-off may be something quite respectable and necessary, but we need to know what it is.

In extreme cases, reductivism finds its pay-off in quite undisciplined, extravagant doctrines, which I have called pieces of Reductive Megalomania. I shall say a little more about these later, but in case anyone doesn't know what I mean, it may be worthwhile just to give a few specimens here.

(1) This world is the will to power and nothing beside . . . Life itself is *essentially* appropriation, injury, subjugation of the strange and the weaker, suppression, severity, imposition of its own forms, incorporation and, at the least and mildest exploitation.
(Friedrich Nietzsche, *The will to power*, para. 1067 and *Beyond good and evil*, para. 269, translated by Marianne Cowan, Gateway, Chicago, 1955. Compare *Beyond good and evil*, para. 36.)

(2) The argument of this book is that we, and all other animals, are machines

created by our genes . . . Like successful Chicago gangsters, our genes have survived, in some cases for millions of years, in a highly competitive world. This entitles us to expect certain qualities in our genes. I shall argue that a predominant quality to be expected in a successful gene is ruthless selfishness . . . If you wish . . . to build a society in which individuals cooperate generously and unselfishly towards a common good, you can expect little help from biological nature. Let us try to teach generosity and altruism, because we are born selfish.
(Richard Dawkins, *The selfish gene*, pp. 2–3. Oxford University Press, 1976.)

(3) Once acclimatized to space-living, it is unlikely that man will stop until he has roamed over and colonized most of the sidereal universe, or that even this will be the end. Man will not ultimately be content to be parasitic on the stars, but will invade them and organize them for his own purposes . . . The stars cannot be allowed to continue in their old way, but will be turned into efficient heat-engines . . . By intelligent organization, the life of the universe could probably be prolonged to many millions of millions of times what it would be without organization.
(J.D. Bernal, *The world, the flesh and the Devil*, pp. 35–6.)

VALUE-FREE?

In such passages, though the tone is sternly reductive, the positive proposals made are obviously extravagances, not economies. Of course not all reductions carry such weird cargo, and you may well wonder whether I am justified in saying—as I do want to say—that reduction is never value-neutral, never just aimed at simplicity, that it is always part of some positive propaganda campaign. You may ask: does reduction always aim in some way to debunk or downgrade what it reduces?

Now certainly the kinds of downgrading involved are very various, and some of them are so mild that they are scarcely noticed. Perhaps the mildest possible kind is the relation between chemistry and physics. Is this perhaps just a formal connection, just a matter of establishing intertranslatability? Does it mark no difference in value between physics and chemistry?

It is quite true that people who point out this relation are not attacking chemistry, nor campaigning to get rid of it. But there *is* still a value-judgement involved here, one of a subtler and more interesting kind. It concerns what is seen as more 'real', more 'fundamental'. Physics wins here because it stands nearest to the end of the quest that dominated science from the time of Galileo until quite lately—the atomistic project of explaining the behaviour of matter by analysing it into solid ultimate particles moved by definite forces—'ultimate building-blocks', as people still revealingly say. Given that quest, and given the faith that it would finally provide the only proper explanation of everything, chemistry inevitably emerged as the subordinate partner, and all other studies as more subordinate still.

Modern chemistry has grown up with that status, and chemists probably don't usually resent it. Some of them have, however, said that this exclusive orientation to physics distorts chemistry. Similar complaints have often been made about distortion of biology, and it is interesting that Francis Crick, himself often a keen reducer, is among those complaining. As he points out, the accumulated effects of evolution give biology a kind of complexity all its own:

All this may make it very difficult for physicists to adapt to most biological research. Physicists are all too apt to look for the wrong sorts of generalizations, to concoct theoretical models that are too neat, too powerful and too clean. Not surprisingly, these seldom fit well with the data. To produce a really good biological theory one must try to see through the clutter produced by evolution to the basic mechanisms lying beneath them . . . What seems to physicists to be a hopelessly complicated process may have been what nature found simplest, because nature could only build on what was already there.
(Francis Crick, *What mad pursuit*, Penguin Books, London, 1989, p. 139.)

We need to ask, then, just what the assumed primacy of physics means. It does not concern only a hierarchy internal to the sciences. Right from the start, physical explanations have been expected to extend very widely, far beyond the boundaries of chemistry. Descartes established the assumption that, since the ultimate particles moved on the model of machines, the physical things made out of them, including human bodies, must do that too. This assumption took the empire of physics right into the realm of human affairs, and, as long as the simple mechanical model held sway, it provided a kind of explanation there that seemed entitled to supersede all other ways of thinking.

Though Descartes himself exempted the mind from this machine, others quickly saw that it too could be reduced to fit the picture. Thomas Hobbes (always a most determined reducer) did this with great zest right across the psychological scene:

When the action of the same object is continued from the Eyes, Ears and other organs to the Heart, *the real effect there is nothing but motion or endeavour*, which consisteth in Appetite or Aversion to or from the object moving. But the appearance, or sense of that motion, is that we either call DELIGHT or TROUBLE OF MIND.
Life itself is but motion.
(*Leviathan*, Part 1, chapter 6. Emphasis mine.)

Physical explanations had primacy because—quite simply—they revealed reality, whereas subjective experience was 'only an appearance'. Reductive psychologizers like Hobbes did not see that there could be objective facts about subjective experience, that an appearance is itself a fact, and that some appearances—for instance the experience of pain or grief, delight or trouble of mind—can be centrally important parts of the facts that affect us. These things do not just appear to matter; they do matter. So we vitally need appropriate conceptual schemes for discussing them.

Hobbes's simple contrast between reality and appearance is easily read as the familiar one between reality and illusion. Ordinary, everyday life is then thought of, in an extraordinarily incoherent way, as some kind of dream or mistake. Thus, Einstein was convinced that ordinary, irreversible time is an illusion, since it had no place in his theory of physics. His close friend Michele Besso tried long and strenuously to convince him that this could not be right, but Einstein remained adamant and when Besso died he wrote, in a letter of condolence to Besso's family, 'Michele has left this strange world just before me. This is of no importance. For us convinced physicists *the distinction between past, present and future is an illusion, though a persistent one.*'[1] (Emphasis mine.) But to say that something is not important does not justify calling it an illusion. Similarly Stephen Hawking (though of course his view of time is different) is happy to say that, 'so-called imaginary time is real time and . . . what we call real time is just a figment of our imaginations'.[2] The term 'imaginary time' is of course supposed to be an inoffensive technical term here. But Hawking shows, by using the ordinary phrase 'a figment of our imaginations' in parallel with it, that he has slipped into using both terms literally.

When people who talk like this are pressed to explain themselves, and are asked whether they really think that the part of our knowledge that falls outside physics—which is almost all our knowledge—is pure illusion, they tend to dither and retreat somewhat. 'No', they say, 'it isn't exactly false or illusory, but it is somehow superficial and provisional. It is a kind of amateur guesswork because it has not yet been properly checked by scientists'. Thus, in a *New Scientist* article, Peter Atkins kindly makes room for other studies, conceding that:

there will always be room for constructing questions that package groups of deeper questions into units appropriate to the level of discourse . . . It *will never be appropriate to exterminate history, law and so on,* any more than it would be appropriate to insist that all discourse in biology should be expressed in terms of particle physics. Concepts must be allowed to operate at a pragmatic level.[3] (Emphasis mine.)

But this, he says, is merely because the deeper scientific account is unfortunately 'too cumbersome for daily use', not because any other enquiry could actually add anything useful to it. Science—meaning essentially just physics and chemistry, since Atkins names biology among the surface layers—remains 'omnicompetent'. It is, he says, able to answer, at the deepest level, all questions that could arise in any enquiry.

TRANSLATION PROSPECTS

Anyone who is tempted by this project should try it out by translating some simple historical statement into the deeper, physical truths that are held to underlie it. What, for instance, about a factual sentence like 'George was allowed home from prison at last on Sunday'? How will the language of physics convey the meaning of 'Sunday'? or 'home' or 'allowed' or 'prison'? or 'at last'? or indeed 'George'? (There are no individuals in physics.) The meaning of all these terms concerns very complex, far-ranging systems of social relation, not the physical details of a particular case.

For a translation, all these social concepts would have to vanish and be represented by terms describing the interactions of groups of particles moved by various forces. The trouble with this new version is *not*, as Atkins says, that it is 'too cumbersome for everyday use', but that it does not begin to convey the meaning of what is said at all. The sentence as it stands does not refer only to the physical items involved. Indeed, most of the physical details are irrelevant to it. (It does not matter, for instance, where the prison is or by what transport or what route George came home.) What the sentence describes is a symbolic transaction between an individual and a huge social background of penal justice, power structures, legislation, and human decisions. The words it uses are suited to fill in that historical and social background. Without such concepts, the whole meaning of the sentence would vanish.

This piece of history—this little narrative sentence—is not something sketchy or provisional. It is not a blueprint that needs scientific validation. It is not emotive expression or amateur 'folk-psychology'. It is solid information of exactly the kind that is needed. It is precise in the way that it needs to be, and if more precision is needed—such as why he is let out—that too can be supplied through concepts of the same kind. And if anyone cares to try the same experiment with a sentence from the law, they will find themselves still more totally baffled.

IDEALIST REDUCTIONS

Reducers feel that there is still something unofficial about this everyday language, because it speaks of entities such as human beings and homes and prisons that are not in the repertoire of physics; entities that cannot, therefore, be quite real. But to question their reality is to invoke not physics but metaphysics. What does 'real' mean here? What deeper reality are we talking about? Why (first) is there this profound faith in extending the mechanical explanatory system indefinitely, and why (next) is that faith expressed in this violently ontological language of appearance, or illusion, and reality?

Plainly, the atomistic pattern of explanation in terms of movements of ultimate, unsplittable particles still has a strong grip on our imaginations. Though physicists no longer believe in those ultimate particles, the kind of simplification which this pattern promised is extremely attractive, and of course it has often worked very well. Besides, from the debunking point of view, reducing wholes to parts is always a good way to downgrade their value. As people say, 'After all, when you get right down to it, a human body is just £5-worth of chemicals . . . .'

There is also, however, a much more serious and reputable wish to bring explanations of mind and matter together in some sort of intelligible relation. Descartes' division of the world between these two superpowers, mind and body, that were scarcely on speaking terms, is most unsatisfactory. Once the pattern of unification by conquest had been proposed as a way of helping thought to cross such gulfs between enquiries, it seemed natural to carry it further. Materialists who reduced mind to matter certainly did think they were simply following the example of physics. But that example cannot decide *which* of the superpowers is to prevail.

Formally, it is just as easy to absorb matter into mind by being sceptical about the existence of outside objects as it is to absorb mind into matter. Phenomenalism works quite as well as materialism. David Hume devised a sceptical, idealist reduction which cut out physical matter as an unnecessary entity just as sternly as it cut out God and the continuing soul. This triumph of general parsimony left Hume with a most obscure world consisting only of particular perceptions: atoms of mind without any real owner. But Hume still thought his economy was based on the example of modern science. In justifying his reduction of all motives to the search for utility, he cited Newton's example, writing hopefully:

It is entirely agreeable to the rules of philosophy, and even of common reason, where any principle has been found to have a great force and energy in one instance, to ascribe to it a like energy in all similar instances. This indeed is Newton's chief rule in philosophizing.
(David Hume, *Enquiry concerning the principles of morals*, Sec. III, part ii, end (para 163).)

This shows most revealingly how tempting it is to see cases as 'similar' once you have got a formula that you hope might fit them all. In defending this fatal tendency, Hume writes, somewhat naïvely, as if simplicity were always just a matter of the actual number of principles mentioned:

Thus we have established two truths without any obstacle or difficulty, that it is from natural principles this variety of causes excite pride and humility, and that it is not by a different principle each cause is adapted to its passion. We shall now proceed to enquire how we may *reduce these principles to a lesser number*, and find among the causes something common on which their influence depends.
(*Treatise of human nature*, Book II, part 1, section 4.)

This hasty habit has been responsible for a whole raft of grossly over-simple reductive theories of motivation. But it did not, of course, settle the question of whether mind should be reduced to matter or matter to mind. Formally, these projects look very similar. Both spring from the strong demand for unity, from the conviction that reality simply cannot be arbitrarily split in two down the middle. But this formal demand for unity cannot help us to pick sides. Nor does it have the sort of force that would be needed to make people go on, as they have, accepting the strange paradoxes that emerge later, when they try to carry through either monolithic materialism or monolithic idealism consistently. At that point, all simplicity is lost. So, if the search for simplicity were the real aim, people would naturally give up the reductive project.

MORAL ASPECTS

What makes theorists carry reduction further must then surely be, not a formal search for order, but the pursuit of an ideal. Though people often claim that their metaphysical views involve no moral bias, metaphysics usually does express some kind of deep attitude to life, and there is no reason on earth why it should not do so. There is nothing disreputable about morals. What is needed is that this attitude should be conscious and openly expressed for discussion. Bias must not be smuggled in as if it were a technical matter only accessible to the experts.

In the case of mind and matter, the moral drama involved is often obvious at an everyday level. For instance, in medicine, and especially in psychiatry, there is often a choice between viewing patients primarily as physical organisms or as conscious agents. As current experience shows, choice can have strong practical consequences for treatment—indeed it can decide the whole fate of the patient. Yet it is often seen as determined abstractly in advance by conceptions of what is scientific.[10]

The metaphysical idea that only the physical body is actually *real*, while talk of the mind or soul is mere superstition, can have a startling influence on conduct here. The opposite folly—of ignoring the body and treating only the mind—of course also has its own metaphysical backing. But that backing tends to be more explicit, and it is not usually a reductive one. Freudian and Existential psychiatrists don't suppose that bodies are actually unreal. They are not Christian Scientists. They don't support their methods by an idealist metaphysic. Hume's path of reductive idealism is in fact too obscure to influence conduct in the way that reductive materialism has.

THOUGHTS AS WHISTLES

This is the asymmetry that makes our current idea of 'reduction' so confusing. It is not, of course, confined to psychiatry. The idea that mind and matter are competitors, that only one of them can determine conduct, has long had a wide influence. Its preferred form today is the 'epiphenomenalist' one devised by T. H. Huxley. This says that consciousness is not exactly unreal, but it is merely surface froth, an ineffectual extra. Our thinking (said Huxley) is like the railway guard's whistle, it makes a lot of noise and may seem to set off the train, but the real cause of locomotion lies in the boiler . . . The body does what it was going to do anyway, and the mind merely paints the scenery for this process, scenery which somehow persuades the owner that he is in charge. Or, as B. F. Skinner put it,

> The punishment of sexual behaviour changes sexual *behaviour*, and any feelings which may arise are at best by-products . . . We do feel certain states of our bodies associated with behaviour, but . . . they are by-products and not to be mistaken for causes . . . Freedom is a matter of contingencies of reinforcement, not of the feelings the contingencies generate.
> (*Beyond freedom and dignity*, Penguin Books, Harmondsworth, 1973, pp. 20–1.)

I can't here go into the fascinating confusions embodied in epiphenomenalism. One interesting question is, of course, how bodies such as Huxley's or Skinner's manage to be so clever as to do all that theorizing if their minds really do not help in the business. Another—which has been recently stressed—is how consciousness could have evolved at all if it really had no kind of effect in the world. Indeed, the idea of anything occurring without having effects is an extremely strange one.[4] Epiphenomenalism is, in fact, one more rather desperate distortion produced by Descartes' violent separation of mind from body. Once these two are seen as totally distinct kinds of item, unable to affect each other, it is never possible to connect them properly again. So reducers repeatedly try to get rid of one party or the other, with results that are never really intelligible.

What is needed has to be something more like a double-aspect account, in which *we do not talk of two different kinds of stuff at all, but of two complementary points of view*: the inner and the outer, subjective and objective. Human beings are highly complex wholes, about which we really don't know very much. We get the partial knowledge that we do have of them in two ways: from the outside and the inside. In general, neither of these ways of knowing has any fixed precedence over the other. They are both useful for different purposes— just as, for instance, sight and touch are useful in different ways for our knowledge of the external world. And, as we mentioned earlier, there are some situations, such as delight, pain, grief, and the like, where the subjective angle is central. Thomas Nagel, in his book *The view from nowhere*, has proposed that relating these viewpoints properly is a central philosophical problem, one which has been distorted repeatedly by various kinds of dualism. He writes:

I want to describe a way of looking at the world and living in it that is suitable for complex beings without a naturally unified standpoint. It is based on a deliberate effort to juxtapose the internal and external or subjective and objective views at full strength, in order to achieve unification when it is possible, and to recognize clearly when it is not. Instead of a unified world view, we get the interplay of these two uneasily related types of conception, and the essentially incompletable effort to reconcile them. (*The view from nowhere*, Oxford University Press, 1986, p. 4.)

MORAL QUESTIONS

On some moral issues, however, there is serious reason for giving precedence to one of these angles. In our tradition, a central motive for materialist reduction has been moral indignation against the Church. Most of the great philosophical reducers have been violently anti-clerical, often with excellent reason. Hobbes's central concern was to discredit the horrible seventeenth-century wars of religion. Hume's was to attack repressive Christian morality, especially the tyranny of the eighteenth-century Calvinistic Scottish church. Nietzsche's was to puncture the complacent sentimentality of his Lutheran upbringing. And so on.

Now atheism and anti-clericalism do not actually require materialism. Athiestic idealism like Hume's is a perfectly possible option and it may be the more coherent one. At the end of the nineteenth century many serious sceptics thought it the clearer choice. (Russell's lifelong ambivalence is quite interesting here.) At present, however, materialism strikes most atheists as a more straightforward path, and it can, of course, more easily tap the prestige of physical science.

Both kinds, however, have converged to suggest that reductivism is primarily a moral campaign against Christianity. This is a dangerous mistake. Obsession with the churches has distracted attention from reduction employed against notions of human individuality, which is now much more damaging. It has also made moral problems look far simpler than they actually are. Indeed, some hopeful humanist reducers still tend to imply that, once Christian structures are cleared away, life in general will be quite all right and philosophy will present no further problems.[5]

In their own times, these anti-clerical reductive campaigns have often been very useful. But circumstances change. New menaces, worse than the one that obsesses us, are always appearing, so that what looked like a universal cure for vice and folly becomes simply irrelevant. In politics, atheistical states, as seen in this century, are not an encouraging omen for the simple secularistic approach to reform. It turns out that the evils that have infested religion are not confined to it, but accompany any successful human institution. Nor is it even clear that religion itself is something that the human race either can or should be cured of.

This kind of atheistic motive for reductivism is, then, something of limited value today, something which needs much more attention and criticism than it often gets. But there are other motives for it, much less noticed, that are really dangerous: primarily, those concerned with the relations between the reducing scientists and the reduced people who are their subject-matter, such as the psychiatric patients just mentioned. When the question is about how a particular person is to be treated, then that person's own viewpoint on the matter has a quite peculiar importance. Psychological theories, like Skinnerian behaviourism, that exclusively exalt the objective standpoint cannot possibly do justice to that importance. Indeed, they exist to bypass it.

Behaviourism was seen as admirably scientific and austere just because it was reductive. Its reductiveness was believed to make it scientifically impartial. Behaviourists dismissed attention to the subjective angle as an irrelevant extravagance, a sentimental luxury that ought to be renounced in the name of science. But this high opinion of its scientific status was not itself a piece of science. It was a propaganda exercise on behalf of a special moral position. The position itself was never defended in the appropriate moral terms, but always as being in some mysterious sense 'scientific'. It remains a highly disturbing piece of dogma.

I suggested earlier that, when we encounter claims to intellectual austerity, such as this one, we should always look for the pay-off. Here, that is not difficult. It is both convenient and flattering for psychologists to regard other people as mechanisms and themselves as the engineers appointed to examine and repair them. To ignore the subject's own views about his or her state naturally makes the work much simpler. It also greatly increases the practitioner's power. No doubt psychologists are often sincere in claiming to act for the good of their 'subjects'—a word with interesting associations. But the principles underlying Skinner's approach simply leave no room for the subjects' own view about what their own good might consist in. Those principles legitimize manipulation unconditionally. No doubt that is one reason why this way of doing psychology is no longer quite so widely favoured as it used to be.

PSYCHOLOGICAL REDUCTION

Psychological reducers, however, have yet another possibility open to them, which can prove even more gratifying. Besides reducing other people's motives to mechanical movements, they can also reduce them to other, underlying, motives which are cruder than those usually admitted. Now the relation between these two enterprises is not clear. If the physicalist reduction works, it is not obvious how the purely psychological one can find room to work as well, or why it is needed. If (for instance) Huxley and Skinner are right to say that all the chapters in this book have just written themselves as a result of a blind movement of particles—that we really didn't have any effective motive for writing them—it is rather hard to see how they have also (actually) been produced by unbridled self-aggrandizement and spite.

Both methods, however, have been practised together since the dawn of modern reduction, without any clear notice of how they should be related. Hobbes constantly uses both. He insists equally that life itself is but motion and that, for instance,

No man giveth, but with intention of Good to himself, because Gift is voluntary, and of all Voluntary acts, the Object is to every man his own Good.
(*Leviathan*, pt. 1, ch. 15.)

Freud also uses both methods, though he prefers the psychological one:

Parental love, which is so moving and at bottom so childish, is nothing but the parents' narcissism born again.
(Paper on 'Narcissism' in Vol. XIV of *Collected works*, p. 91.)

and so on . . . Now I don't at all want to say that this kind of diagnosis of underlying motives is always wrong or unjustified. On the contrary, I think it is often called for, and it can be practised responsibly to great profit. But I do want to point out the enormous inducements there are to practise it indiscriminately and wildly instead. The pleasure of showing other people up as moral frauds, combined with the intellectual satisfaction of extending one's guiding theory more and more widely, is a pay-off that theorists find it very hard to resist.

Thus, even straightforward psychological reduction of motives is a tricky business, by no means always austere and well-guided. But what is far worse is a bastard mixture of this with the physicalist reduction, losing all the advantages of both. This muddle is chronic among sociobiologists. Thus David Barash, updating Freud:

Parental love itself is but an evolutionary strategy whereby genes replicate themselves . . . We will analyse parental behaviours, the underlying selfishness of our behaviour to others, even our own children.
(*Sociobiology, the whisperings within*, p. 3.)

Human behaviour—like the deepest capacities for emotional response which drive and guide it—is the circuitous technique by which human genetic material has been and will be kept intact. Morality has no other demonstrable ultimate function.
(Edward O. Wilson, *On human nature*, p. 167.)

The evolutionary theory of human altruism is greatly complicated by the ultimately self-serving quality of most forms of that altruism. No sustained form of human altruism is explicitly and totally self-annihilating. Lives of the most towering heroism are paid out in the expectation of great reward, not the least of which is a belief in personal immortality.

Compassion is selective and often self-serving . . . it conforms to the best interests of self, family and allies of the moment.
(Edward O. Wilson, *On human nature*, pp.154–5.)

Officially, sociobiology doesn't say anything about motives at all. Its business is only with the statistical probability that certain types of action will affect the future distribution of an agent's genes. Disastrously, however, sociobiologists have chosen to describe this harmless topic in the language of motive, using words like 'selfishness', 'spite', and 'altruism' as technical terms for various distributive tendencies. (Both this passage from Barash and the one from Dawkins quoted earlier, by the way, occur in the first pages of their books, *before* the special, technical definition of selfishness has been explained.)

Because this language of motive is so natural and habitual in its ordinary sense, these authors constantly slip into mixing the two systems and thus supposing that they have radically explained human psychology. Dreams of still wider academic empire, involving the reductive conquest of all other studies, naturally follow:

It may not be too much to say that sociology and the other social sciences, as well as the humanities, are the last branches of biology waiting to be included in the Modern Synthesis. One of the functions of Sociobiology, [just *one*] then, is to reformulate the foundations of the social sciences in a way that draws these subjects into the Modern Synthesis.

Stress will be evaluated in terms of the neurophysiological perturbations and their relaxation times. Cognition will be translated into circuitry. Learning and creativeness will be defined as the alteration of specific portions of the cognitive machinery regulated by input from the emotive centres. Having cannibalized psychology, the new neurobiology will yield an enduring set of first principles for sociology.
(Edward O. Wilson, *Sociobiology*, pp. 4 and 575.)

None of this would have looked even faintly plausible if a hasty combination of physicalist and psychological reductions had not given these writers the impression that they had finally summed up human psychology. If passages like these don't constitute megalomania, I don't know what does. About them, I rest my case.

The other region of fantasy that I have lately been exploring is the range of predictions now being made, not just by ordinary prophets but by eminent scientists, about an eventual human conquest of the whole universe.[6] Essentially, these predictions are Lamarckian, extrapolating what is seen as a rising graph of evolution that will exalt the human race to the skies, giving it an increasingly glorious, and perhaps unending, future.[7] These predictions postulate highly successful space travel and an even more successful transfer of human consciousness (whatever that is) to machines. They have no support from current Darwinian biology, which completely rejects the Lamarckian graph. Instead they rest on highly abstract arguments drawn from cosmology, from dubious probability theory, and from certain areas of artificial intelligence.

The prophecies were first made half a century back by J. B. S. Haldane and J. D. Bernal, two very distinguished and imaginative scientists who were devout dialectical materialists. Marxism had accustomed both of them to debunking everyday concepts by sternly reductive rhetoric, and also to using the promises of a remote and splendid future in order to justify ignoring the crimes and miseries of the present. These two unlucky habits are surely what betrayed them into their compensatory dreams of a distant future. Bernal's little book has been admiringly quoted by several modern proponents of this compensatory myth.[8] It combines strongly the two elements I have been noting: harsh, austere contempt for ordinary ways of thinking, and unbridled indulgence in power-fantasies. In particular, Bernal shows an extraordinary, paranoid revulsion from the human body:

Modern mechanical and modern biochemical discoveries have rendered both the skeletal and metabolic functions of the body to a great extent useless . . . Viewed from the standpoint of the mental activity by which he (man) increasingly lives, it is a highly inefficient way of keeping his mind working. In a civilised worker, the limbs are mere parasites, demanding nine-tenths of the energy of the food and even a kind of blackmail in the exercise they need in order to prevent disease, while the body organs wear themselves out in supplying their requirements . . . Sooner or later the useless parts of the body must be given more modern functions or dispensed with altogether. (*The world, the flesh and the Devil*, pp. 41–2[9].)

Is that austere enough for us? I can't take time here to examine the modern versions of this story. I have quoted Bernal because his forceful style shows so clearly the strange ambivalences of reductive rhetoric. His chilling, deadpan tone crushes diffident readers into accepting his openly ludicrous visions as if they were sober, practical proposals. I hope this passage may serve to point out what I chiefly want to say: namely that, though reduction is often a useful tool, not all reductions are either useful or sensible.

REFERENCES

1. *Correspondence, Albert Einstein–Michele Besso, 1903–1955*. Herman, Paris, 1972. Quoted by Ilya Prigogine and Isabelle Stengers in *Order out of chaos; Man's new dialogue with Nature*, p. 294. Collins, Fontana, London, 1985.
2. Stephen Hawking, *A brief history of time*, p. 139. Bantam Press, London and New York, 1988.
3. Will science ever fail? *New Scientist*, 8 August 1992, answering an article of mine, strangely titled by the editor 'Can science save its soul?', *New Scientist*, 1 August 1992.
4. I have discussed the roots of these strange beliefs, with other aspects of reductionism, in *The ethical primate; humans, morality and freedom*. Routledge, London, 1994–5 (in press).
5. See, for instance, the extraordinarily optimistic prophecies at the ends of Edward O. Wilson's two books *Sociobiology, the modern synthesis*, p. 574 (Belknap Press, Harvard, 1975) and *On human nature*, pp. 201–9 (Harvard University Press, 1978).
6. In *Science as salvation; a modern myth and its meaning*. Routledge, London, 1992.
7. I explored a wider field of these predictions—not specially concerned with space— in *Evolution as a religion: strange hopes and stranger fears*. Methuen, London, 1985.
8. For instance by John D. Barrow and Frank R. Tipler in their remarkable tome *The anthropic cosmological principle*, Oxford University Press, 1986, pp. 618–19.
9. Cape, London, 1929; reprinted Indiana University Press, Bloomington, 1989. For Haldane's contribution see his *Possible worlds* (Chatto and Windus, London, 1927, p. 287).
10. See *Psychiatric polarities, methodology, and practice* (ed. P. R. Slavney and P. R. McHugh). Johns Hopkins University Press, 1987.

CHAPTER TEN

Artificial intelligence and human dignity

MARGARET A. BODEN

THE PROBLEM—AND WHY IT MATTERS

Let us begin with a quotation—or, rather, several:

For the many, there is hardly concealed discontent . . . 'I'm a machine,' says the spot welder. 'I'm caged,' says the bank teller, and echoes the hotel clerk. 'I'm a mule,' says the steel worker. 'A monkey can do what I can do,' says the receptionist. 'I'm less than a farm implement,' says the migrant worker. 'I'm an object,' says the high-fashion model. Blue collar and white call upon the identical phrase: 'I'm a robot.'
(Terkel 1974, p. xi.)

Studs Terkel encountered these remarks during his study of American attitudes to employment. What relevance can they have here? Welders and fashion models are not best known for an interest in philosophy. Blue collar and white, surely, have scant interest in the abstract issue of scientific reductionism?

The 'surely', here, is suspect. Admittedly, neither the blue nor the white feel much at ease with philosophical terminology. But ignorance of jargon does not imply innocence of issues.

These workers clearly took for granted, as most people do, that there is a clear distinction between humans on the one hand and animals—and machines—on the other. They took for granted, too, that this distinction is grounded in the variety of human skills and, above all, in personal autonomy. When their working conditions gave no scope for their skills and autonomy, they experienced not merely frustration but also personal threat—not least, to their sense of worth, or human dignity.

'So much the worse for them, poor deluded fools!', some might retort, appealing not only to (scientific) truth but also to what they see as (humanistic) illusion—specifically, the illusion of freedom inherent in the notion of human dignity.

The behaviourist B. F. Skinner, for example, argued that 'the literature of dignity . . . stands in the way of further human achievements' (Skinner 1971, p. 59), the main achievement he had in mind being the scientific understanding of human behaviour. 'Dignity', he said, 'is a matter of giving people credit, of admiring them for their (self-generated) achievements.' But his behaviourist principles implied that 'the environment', not 'autonomous man', is in control (Skinner 1971, p. 21). No credit, then, to *us*, if we exercise some skill—whether bodily, mental, or moral. Spot welder and fashion model can no longer glory in their dexterity or gracefulness, nor clerk and cleric in their profession or vocation. Honesty and honest toil alike are de-credited, de-dignified.

Behaviourism, then, questions our notions of human worth. But it is at least concerned with life. Animals are living things, and *Rattus norvegicus* a moderately merry mammal. Some small shred of our self-respect can perhaps be retained, if we are classed with rats, or even pigeons. But artificial intelligence, it seems, is another matter. For AI compares us with computers, and dead, automatic tin-cannery is all they are capable of. Sequential or connectionist, it makes no difference: machines are not even alive. The notion that they could help us to an adequate account of the mind seems quite absurd.

The absurdity is compounded with threat. For (on this view) it seems that if human minds were understood in AI terms, everything we think of as distinctively human—freedom, creativity, morals—would be explained away. Ultimately, a computational psychology and neuroscience would reduce these matters to a set of chemical reactions and electrical pulses. No autonomy there . . . and no dignity, either. We could not exalt human skills and personality above the dexterity of monkeys or the obstinacy of mules. As for honouring excellence in the human mind, this would be like preferring a Rolls-Royce to a Mini: some basis in objectivity, no doubt, but merely a matter of ranking machines.

Given these widespread philosophical assumptions, it is no wonder if AI is feared by ordinary people. They think of it as even more threatening to their sense of personal worth than either industrial automation or 'mechanical' work-practices, the subjects of the complaints voiced to Terkel.

What they think *matters*. Given the central constructive role in our personal life of the self-concept, we should expect that people who believe (or even half-believe) they are mere machines may behave accordingly. Similarly, people who feel they are being treated like machines, or caged animals, may be not only frustrated and insulted but also insidiously lessened by the experience.

Such malign effects can indeed be seen, for instance in psychotherapists' consulting rooms. Thirty years ago, before the general public had even heard of AI, the therapist Rollo May remarked on some depersonalizing effects of behaviourism, and of reductionist science in general:

I take very seriously . . . the dehumanizing dangers in our tendency in modern science to make man over into the image of the machine, into the image of the techniques by which we study him . . . A central core of modern man's 'neurosis' is the undermining of his experience of himself as responsible, the sapping of his willing and decision. (May 1961, p. 20.)

I have used this quote elsewhere, but make no apology for repeating it. It shows the practical results of people's defining themselves as (what they think of as) machines, not only in a felt unhappiness but also in an observable decline of personal autonomy.

The upshot is that it is practically important, not just theoretically interesting, to examine the layman's philosophical assumptions listed above. Are they correct? Or are they mere sentimental illusion, a pusillanimous refusal to face scientific reality? In particular, are AI concepts and AI explanations compatible with the notion of human dignity?

AI AND ANTS

At first sight, the answer may appear to be 'No'. For it is not only behaviourists who see conditions in the external environment as causing apparently autonomous behaviour. Only a few years after May's complaint quoted above, Herbert Simon—a founding-father of AI—took much the same view (Simon 1969).

Simon described the erratic path of the ant, as it avoided the obstacles on its way to nest or food, as the result of a series of simple and immediate reactions to the local details of the terrain. He did not stop with ants, but tackled humans too. For over twenty years, Simon has argued that rational thought and skilled behaviour are largely triggered by specific environmental cues. The extensive psychological experiments and computer modelling on which his argument is based were concerned with chess, arithmetic, and typing (Newell and Simon 1972; Card *et al.* 1983). But he would say the same of bank-telling and spot-welding.

Simon's ant was not taken as a model by most of his AI colleagues. Instead, they were inspired by his earliest, and significantly different, work on the computer simulation of problem-solving (Newell and Simon 1961; Newell *et al.* 1963). This ground-breaking theoretical research paid no attention to environmental factors, but conceived of human thought in terms of internal mental/computational processes, such as hierarchical means–end planning and goal-representations.

Driven by this 'internalist' view, the young AI community designed—and in some cases built—robots guided top-down by increasingly sophisticated internal planning and representation (Boden 1987, ch. 12). Plans were worked out ahead of time. In the most flexible cases, certain contingencies could be foreseen, and the detailed movements, and even the sub-plans, could be decided on at the time of execution. But even though they inhabited the physical world, these robots were not real-world, real-time, creatures. Their environments were simple, highly predictable, 'toy-worlds'. They typically involved a flat ground-plane, polyhedral and/or pre-modelled shapes, white surfaces, shadowless lighting, and—by human standards—painfully slow movements. Moreover, they were easily brought to a halt, or trapped into fruitless perseverative behaviour, by unforeseen environmental details.

Recently, however, the AI pendulum has swung towards the ant. Current research in *situated robotics* sees no need for the symbolic representations and detailed anticipatory planning typical of earlier AI robotics. Indeed, the earlier strategy is seen as not just unnecessary, but ineffective. Traditional robotics suffers from the brittleness of classical AI programs in general: unexpected input can cause the system to do something highly inappropriate, and there is no way in which the problem-environment can help guide it back on to the right track. Accepting that the environment cannot be anticipated in detail, workers in situated robotics have resurrected the insight—often voiced within classical AI, but also often forgotten—that the best source of information about the real world is the real world itself.

Accordingly, the 'intelligence' of these very recent robots is in the hardware, not the software (Braitenberg 1984; Brooks 1991). There is no high-level program doing detailed anticipatory planning. Instead, the creature is engineered in such a way that, within limits, it naturally does the right (adaptive) thing at the right time. Behaviour apparently guided by goals and hierarchical planning can, nevertheless, occur (Maes 1991).

Situated robotics is closely related to two other recent forms of computer modelling, likewise engaged in studying 'emergent' behaviours. These are *genetic algorithms* (GAs) and *artificial life* (A-life).

GA-systems are self-modifying programs, which continually come up with new rules (new structures) (Holland 1975; Holland *et al.* 1986). They use rule-changing algorithms modelled on genetic processes such as mutation and crossover, and algorithms for identifying and selecting the relatively successful rules. Mutation makes a change in a single rule; crossover brings about a mix of two, so that (for instance) the left-hand portion of one rule is combined with the right-hand portion of the other. Together, these algorithms (working in parallel) generate a new system better adapted to the task in hand.

One example of a GA-system is a computer-graphics program written by Karl Sims (1991). This program uses genetic algorithms to generate new images, or patterns, from pre-existing images. Unlike most GA-systems, the selection of the 'fittest' examples is not automatic, but is done by the programmer—or by someone fortunate enough to be visiting his office while the program is being run. That is, the human being selects the images which are aesthetically pleasing, or otherwise interesting, and these are used to 'breed' the next generation. (Sims could provide automatic selection rules, but has not yet done so—not only because of the difficulty of defining aesthetic criteria, but also because he aims to provide an interactive graphics environment, in which human and computer can cooperate in generating otherwise unimaginable images.)

In a typical run of the program, the first image is generated at random (but Sims can feed in a real image, such as a picture of a face, if he wishes). Then the program makes nineteen independent changes (mutations) in the initial image-generating rule, so as to cover the VDU-screen with twenty images: the first, plus its nineteen ('asexually' reproduced) offspring. At this point, the human uses the computer mouse to choose either *one* image to be mutated, or *two* images to be 'mated' (through crossover). The result is another screenful of twenty images, of which all but one (or two) are newly generated by random mutations or crossovers. The process is then repeated, for as many generations as one wants.

The details of this GA-system need not concern us. However, so as to distinguish it from magic, a few remarks may be helpful. It starts with a list of twenty very simple LISP-functions. A 'function' is not an actual instruction, but an instruction-schema: more like $x + y$ than $2 + 3$. Some of these functions can alter parameters in pre-existing functions: for example, they can divide or multiply numbers, transform vectors, or define the sines or cosines of angles. Some can combine two pre-existing functions, or nest one function inside another (so multiply nested hierarchies can eventually result). A few are basic image-generating functions, capable (for example) of generating an image consisting of vertical stripes. Others can process a pre-existing image, for instance by altering the light-contrasts so as to make 'lines' or 'surface-edges' more or less visible. When the program chooses a function at random, it also randomly chooses any missing parts. So if it decides to *add* something to an existing number (such as a numerical parameter inside an image-generating function), and the 'something' has not been specified, it randomly chooses the amount to be added. Similarly, if it decides to *combine* the pre-existing function with some other function, it may choose that function at random.

As for A-life, this is the attempt to discover the abstract functional principles underlying life in general (Langton 1989). A-life is closely related to AI (and uses various methods which are also employed in AI). One might define A-life as the abstract study of life, and AI as the abstract study of mind. But if one assumes that life prefigures mind, that cognition is—and must be—grounded in self-organizing adaptive systems, then the whole of AI may be seen as a subclass of A-life. Work in A-life is therefore potentially relevant to the question of how AI relates to human dignity.

Research in A-life uses computer modelling to study processes that start with relatively simple, locally interacting units, and generate complex individual and/or group behaviours. Examples of such behaviours include self-organization, reproduction, adaptation, purposiveness, and evolution.

Self-organization is shown, for instance, in the flocking behaviour of flocks of birds, herds of cattle, and schools of fish. The entire group of animals seems to behave as one unit. It maintains its coherence despite changes in direction, the (temporary) separation of stragglers, and the occurrence of obstacles— which the flock either avoids or 'flows around'. Yet there is no overall director working out the plan, no sergeant-major yelling instructions to all the individual animals, and no reason to think that any one animal is aware of the group as a whole. The question arises, then, how behaviour of this kind is possible.

Ethologists argue that communal behaviour of large groups of animals must depend on local communications between neighbouring individuals, who have no conception of the group-behaviour as such. But just what are these 'local communications'?

Flocking has been modelled within A-life, in terms of a collection of very simple units, called Boids (Reynolds 1987). Each Boid follows three rules: (1) keep a minimum distance from other objects, including other Boids; (2) match velocity to the average velocity of the Boids in the immediate neighbourhood; (3) move towards the perceived centre of mass of the Boids in the neighbourhood. These rules, depending as they do only on very limited, local, information, result in the holistic flocking behaviour just described. It does not follow, of course, that real birds follow just those rules: that must be tested by ethological studies. But this research shows that it is at least *possible* for group behaviour of this kind to depend on very simple, strictly local, rules.

Situated robotics, GAs, and A-life could be combined, for they share an emphasis on bottom-up, self-adaptive, parallel processing. At present, situated robots are hand-crafted. In principle, they could be 'designed' by evolutionary algorithms from the GA/A-life stable. Simulated (unembodied) robots have already been 'evolved' by A-life means and the automatic evolution of real physical robots has just begun (Brooks 1992; Cliff *et al.* 1993).

These three research fields have strong links with biology: with neuro-science, ethology, genetics, and the theory of evolution. As a result, animals are becoming theoretically assimilated to *animats* (Meyer and Wilson 1991). The behaviour of swarms of bees and of ant-colonies is hotly discussed at A-life conferences, and entomologists are constantly cited in the A-life and situated-robotics literatures (Lestel 1992). Environmentally situated (and formally defined) accounts of apparently goal-seeking behaviour in various animals, including birds and mammals, are given by (some) ethologists (McFarland 1989). And details of invertebrate psychology, such as visual tracking in the hoverfly, are modelled by research in connectionist AI (Cliff 1991, 1992).

In short, Simon's ant is now sharing the limelight on the AI stage. Some current AI is more concerned with artificial insects than with artificial human minds. But—what is of particular interest to us here—this form of AI sees itself as designing 'autonomous agents' (as A-life in general seeks to design 'autonomous systems').

AUTONOMOUS AGENCY

Autonomy is ascribed to these artificial insects because it is their intrinsic physical structure, adapted as it is to the kinds of environmental problem they are likely to meet, which enables them to act appropriately. Unlike traditional robots, their behaviour is not directed by complex software written for a general-purpose machine, imposed on their bodies by some alien (human) hand. Rather, they are specifically constructed to adapt to the particular environment they inhabit.

We are faced, then, with two opposing intuitions concerning autonomy. Our (and Skinner's) original intuition was that response determined by the external environment lessens one's autonomy. But the *nouvelle*-AI intuition is that to be in thrall to an internal plan is to be a mere puppet. (Notice that one can no longer say 'a mere robot'.) How can these contrasting intuitions be reconciled?

Autonomy is not an all-or-nothing property. It has several dimensions, and many gradations. Three aspects of behaviour—or rather, of its control—are crucial. First, the extent to which response to the environment is direct (determined only by the present state in the external world) or indirect (mediated by inner mechanisms partly dependent on the creature's previous history). Second, the extent to which the controlling mechanisms were self-generated rather than externally imposed. And third, the extent to which inner directing mechanisms can be reflected upon, and/or selectively modified in the light of general interests or the particularities of the current problem in its environmental context. An individual's autonomy is the greater, the more its behaviour is directed by self-generated (and idiosyncratic) inner mechanisms, nicely responsive to the specific problem-situation, yet reflexively modifiable by wider concerns.

The first aspect of autonomy involves behaviour mediated, in part, by inner mechanisms shaped by the creature's past experience. These mechanisms may, but need not, include explicit representations of current or future states. It is controversial, in ethology as in philosophy, whether animals have explicit internal representations of goals (Montefiore and Noble 1989). And, as we have seen, AI includes strong research programmes on both sides of this methodological fence. But this controversy is irrelevant here. The important distinction is between a response wholly dependent on the current environmental state (given the original, 'innate', bodily mechanisms), and one largely influenced by the creature's experience. The more a creature's past experience differs from that of other creatures, the more 'individual' its behaviour will appear.

The second aspect of autonomy, the extent to which the controlling mechanisms were self-generated rather than externally imposed, may seem to be the same as the first. After all, a mechanism shaped by experience is sensitive to the past of that particular individual—which may be very different from that of other, initially comparable, individuals. But the distinction, here, is between behaviour that 'emerges' as a result of self-organizing processes, and behaviour that was deliberately prefigured in the design of the experiencing creature.

In computer-simulation studies within A-life, and within situated robotics also, holistic behaviour—often of an unexpected sort—may emerge. It results, of course, from the initial list of simple rules concerning locally interacting units. But it was neither specifically mentioned in those rules, nor (often) foreseen when they were written.

A flock, for example, is a holistic phenomenon. A birdwatcher sees a flock of birds as a unit, in the sense that it shows behaviour that can be described only at the level of the flock itself. For instance, when it comes to an obstacle, such as a tall building, the flock divides and 'flows' smoothly around it, reorganizing itself into a single unit on the far side. But no individual bird is divided in half by the building. And no bird has any notion of the flock as a whole, still less any goal of reconstituting it after its division.

Clearly, flocking behaviour must be described on its own level, even though it can be explained by (reduced to) processes on a lower level. This point is especially important if 'emergence-hierarchies' evolve as a result of new forms of perception, capable of detecting the emergent phenomena *as such*. Once a holistic behaviour has emerged it, or its effects, may be detected (perceived) by some creature or other—including, sometimes, the 'unit-creatures' making it up.

This implies that a creature's perceptual capacities cannot be fully itemized for all time. In Gibsonian terms, one might say that evolution does not know what all the relevant affordances will turn out to be, so cannot know how they will be detected. The current methodology of AI and A-life does not allow for 'latent' perceptual powers, actualized only by newly emerged environmental features. This is one of the ways in which today's computer modelling is biologically unrealistic (Kugler, 1992).

If the emergent phenomenon can be detected, it can feature in rules governing the perceiver's behaviour. Holistic phenomena on a higher level may then result . . . and so on. Ethologists, A-life workers, and situated roboticists all assume that increasingly complex hierarchical behaviour can arise in this sort of way. The more levels in the hierarchy, the less direct the influence of environmental stimuli—and the greater the behavioural autonomy.

Even if we can *explain* a case of emergence, however, we cannot necessarily *understand* it. One might speak of intelligible vs. unintelligible emergence.

Flocking gives us an example of the former. Once we know the three rules governing the behaviour of each individual Boid, we can see lucidly how it is that holistic flocking results.

Sims's computer-generated images give us an example of the latter. One may not be able to say just why *this* image resulted from *that* LISP-expression. Sims himself cannot always explain the changes he sees appearing on the screen before him, even though he can access the mini-program responsible for any image he cares to investigate, and for its parent(s) too. Often, he cannot even 'genetically engineer' the underlying LISP-expression so as to get a particular visual effect. To be sure, this is partly because his system makes several changes simultaneously, with every new generation. If he were to restrict it to making only one change and studied the results systematically, he could work out just what was happening. But when several changes are made in parallel, it is often impossible to understand the generation of the image *even though* the 'explanation' is available.

Where real creatures are concerned, of course, we have multiple interacting changes, and no explanation at our finger-tips. At the genetic level, these multiple changes and simultaneous influences arise from mutations and crossover. At the psychological level, they arise from the plethora of ideas within the mind. Think of the many different thoughts that arise in your consciousness, more or less fleeting, when you face a difficult choice or moral dilemma. Consider the likelihood that many more conceptual associations are being activated unconsciously in your memory, influencing your conscious musings accordingly. Even if we had a listing of all these 'explanatory' influences, we might be in much the same position as Sims, staring in wonder at one of his *n*th-generation images and unable to say why *this* LISP-expression gave rise to it. In fact, we cannot hope to know about more than a fraction of the ideas aroused in human minds (one's own, or someone else's) when such choices are faced.

The third criterion of autonomy listed above was the extent to which a system's inner directing mechanisms can be reflected upon, and/or selectively modified, by the individual concerned. One way in which a system can adapt its own processes, selecting the most fruitful modifications, is to use an 'evolutionary' strategy such as the genetic algorithms mentioned above. It may be that something broadly similar goes on in human minds. But the mutations and selections carried out by GAs are modelled on biological evolution, not conscious reflection and self-modification. And it is conscious deliberation that many people assume to be the crux of human autonomy.

For the sake of argument, let us accept this assumption at face value. Let us ignore the mounting evidence, from Freud to social psychology (e.g. Nisbett and Ross 1980), that our conscious thoughts are less relevant than we like to think. Let us ignore neuroscientists' doubts about whether our conscious intentions actually direct our behaviour (as the folk-psychology of 'action' assumes) (Libet 1987). Let us even ignore the fact that *unthinking spontaneity*—the opposite of conscious reflection—is often taken as a sign of individual freedom. (Spontaneity may be based in the sort of multiple constraint satisfaction modelled by connectionist AI, where many of the constraints are drawn from the person's idiosyncratic experience.) What do AI, and AI-influenced psychology, have to say about conscious thinking and deliberate self-control?

Surprisingly, perhaps, the most biologically realistic (more accurately: the least biologically unrealistic) forms of AI cannot help us here. Ants, and artificial ants, are irrelevant. Nor can connectionism help. It is widely agreed, even by connectionists, that conscious thought requires a sequential 'virtual machine', more like a von Neumann computer than a parallel-processing neural net. As yet, we have only very sketchy ideas about how the types of problem-solving best suited to conscious deliberation might be implemented in connectionist systems.

The most helpful AI approach so far, where conscious deliberation is involved, is GOFAI: good old-fashioned AI (Haugeland 1985)—much of which was inspired by human introspection. Consciousness involves reflection on one level of processes going on at a lower level. Work in classical AI, such as the work on planning mentioned above, has studied multi-level problem-solving. Computationally informed work in developmental psychology has suggested that flexible self-control, and eventually consciousness, result from a series of 'representational redescriptions' of lower-level skills (Clark and Karmiloff-Smith 1993).

Representational redescriptions, many-levelled maps of the mind, are crucial to creativity (Boden 1990, esp. ch. 4). Creativity is an aspect of human autonomy. Many of Terkel's workers were frustrated because their jobs allowed them no room for creative ingenuity. Our ability to think new thoughts in new ways is one of our most salient, and most valued, characteristics.

This ability involves someone's doing something which they not only *did not* do before, but which they *could not* have done before. To do this, they must either explore a formerly unrecognized area of some pre-existing 'conceptual space', or transform some dimension of that generative space. Transforming the space allows novel mental structures to arise that simply could not have been generated from the initial set of constraints. The nature of the creative novelties depends on which feature has been transformed, and how. Conceptual spaces, and procedures for transforming them, can be clarified by thinking of them in computational terms. But this does not mean that creativity is predictable, or even fully explicable *post hoc*: for various reasons (including those mentioned above), it is neither (Boden 1990, ch. 9).

Autonomy in general is commonly associated with unpredictability. Many people feel AI to be a threat to their self-esteem because they assume that it involves a deterministic predictability. But they are mistaken. Some connectionist AI systems include non-deterministic (stochastic) processes, and are more efficient as a result.

Moreover, determinism does not always imply predictability. Workers in A-life, for instance, justify their use of computer simulation by citing chaos theory, according to which a fully deterministic dynamic process may be theoretically unpredictable (Langton 1989). If there is no analytic solution to the differential equations describing the changes concerned, the process must simply be 'run', and observed, to know what its implications are. The same is true of many human choices. We cannot always predict what a person will do. Moreover, predicting *one's own* choices is not always possible. One may have to 'run one's own equations' to find out what one will do, since the outcome cannot be known until the choice is actually made.

CONCLUSION

One of the pioneers of A-life has said: 'The field of Artificial Life is unabashedly mechanistic and reductionist. However, this *new mechanism*—based as it is on multiplicities of machines and on recent results in the fields or nonlinear dynamics, chaos theory, and the formal theory of computation—is vastly different from the mechanism of the last century'. (Langton 1989, p.6; italics in original).

Our discussion of A-life and *nouvelle* AI has suggested just how vast this difference is. Similarly, the potentialities of classical AI systems go far beyond what most people—fashion-models, spot-welders, bank-tellers—think of as 'machines'. If this is reductionism, it is very different from the sort of reductionism which insists that the only scientifically respectable concepts lie at the most basic ontological level (neurones and biochemical processes, or even electrons, mesons, and quarks).

In sum, AI does not reduce our respect for human minds. If anything, it increases it. Far from denying human autonomy, it helps us to understand how it is possible. The autonomy of Terkel's informants was indeed compromised—but by inhuman working conditions, not by science. Science in general, and AI in particular, need not destroy our sense of human dignity.

REFERENCES

Boden, M.A. (1987). *Artificial intelligence and natural man* (2nd edn). MIT Press, London.

Boden. M.A. (1990). *The creative mind: myths and mechanisms.* Weidenfeld and Nicolson, London.

Braitenberg, V. (1984). *Vehicles: essays in synthetic psychology.* MIT Press, Cambridge, Mass.

Brooks, R.A. (1991). Intelligence without representation. *Artificial Intelligence,* **47,** 139–59.

Brooks, R.A. (1992). Artificial life and real robots. In *Toward a practice of autonomous systems: proceedings of the first European conference on artificial life* (ed. F.J. Varela and P. Bourgine), pp. 3–10. MIT Press, Cambridge, Mass.

Card, S.K., Moran, T.P., and Newell, A. (1983). *The psychology of human–computer interaction.* Erlbaum, Hillsdale, New Jersey.

Clark, A. and Karmiloff-Smith, A. (1993). The cognizer's innards, *Mind and Language,* **8,** 487–515.

Cliff, D. (1990). The computational hoverfly: a study in computational neuroethology. In *From animals to animats: proceedings of the first international conference on simulation of adaptive behaviour* (ed. J.-A. Meyer and S.W. Wilson), pp. 87–96, MIT Press, Cambridge, Mass.

Cliff, D. (1992). Neural networks for visual tracking in an artificial fly. In *Toward a practice of autonomous systems: proceedings of the first European conference on artificial life* (ed. F.J. Varela and P. Bourgine), pp. 78–87, MIT Press, Cambridge, Mass.

Cliff, D. Harvey, I., and Husbands, P. (1993). Explorations in evolutionary robotics. *Adaptive Behaviour,* **2,** 73–110.

Haugeland, J. (1985). *Artificial intelligence: the very idea*. MIT Press, Cambridge, Mass.

Holland, J.H. (1975). *Adaptation in natural and artificial systems: an introductory analysis with applications to biology, control, and artificial intelligence*. University of Michigan Press, Ann Arbor. (Reissued, MIT Press, Cambridge, Mass., 1991.)

Holland, J.H., Holyoak, K.J., Nisbet, R.E., and Thagard, P.R. (1986). *Induction: processes of inference, learning, and discovery*. MIT Press, Cambridge, Mass.

Kugler. (1992). Talk given at the Summer School on 'Comparative approaches to cognitive science', Aix-en-Provence (Organizers, J.-A. Meyer and H.L. Roitblat).

Langton, C.G. (1989). Artificial life. In *Artificial life: proceedings of an interdisciplinary workshop on the synthesis and simulation of living systems* (ed. C. Langton), pp. 1–47. Addison-Wesley, New York.

Lestel, D. (1992). Fourmis cybernétiques et robots-insectes: Socialité et cognition à l'interface de la robotique et de l'ethologie expérimentale. *Information sur les sciences sociales*, **31**(2), 179–211.

Libet, B. (1987). Are the mental experiences of will and self-control significant for the performance of a voluntary act? *Behavioral and Brain Sciences*, **10**, 783–86.

Maes, P. (ed.) (1991). *Designing autonomous agents*. MIT Press, Cambridge, Mass.

May, R. (1961). *Existential psychology*. Random House, New York.

McFarland, D. (1989). Goals, no-goals, and own-goals. In *Goals, no-goals, and own-goals* (ed. A. Montefiore and D. Noble), pp. 39–57. Unwin Hyman, London.

Meyer, J.-A. and Wilson, S.W. (ed.) (1991). *From animals to Animats: proceedings of the first international conference on simulation of adaptive behaviour*. MIT Press, Cambridge, Mass.

Montefiore, A. and Noble, D. (ed.) (1989). *Goals, no-goals, and own-goals*. Unwin Hyman, London.

Newell, A. and Simon, H.A. (1961). GPS–a program that simulates human thought. In *Lernende automaten* (ed. H. Billing), pp. 109–24. Oldenbourg, Munich. Reprinted in *Computer and thought* (ed. E.A. Feigenbaum and J. Feldman), pp. 279–96. McGraw-Hill, New York (1963).

Newell, A. and Simon, H.A. (1972). *Human problem solving*. Prentice-Hall, Englewood Cliffs, NJ.

Newell, A., Shaw, J.C., and Simon, H.A. (1963). Empirical explorations with the logic theory machine: a case-study in heuristics. In *Computers and thought* (ed. A. Feigenbaum and J. Feldman), pp. 109–33. McGraw-Hill, New York.

Nisbett, R.E. and Ross, L. (1980). *Human inference: strategies and shortcomings in social judgment*. Prentice-Hall, Englewood Cliffs, NJ.

Reynolds, C.W. (1987). Flocks, herds, and schools: a distributed behavioral model. *Computer Graphics*, **21**(4), 25–34.

Simon, H.A. (1969). *The sciences of the artificial*. MIT Press, Cambridge, Mass.

Sims, K. (1991). Artificial evolution for computer graphics. *Computer graphics*, **25**(4), 319–28.

Skinner, B.F. (1971). *Beyond freedom and dignity*. Alfred Knopf, New York.

Terkel, S. (1974). *Working*. Pantheon, New York.

CHAPTER ELEVEN

On 'computabilism' and physicalism: some subproblems

HAO WANG

Thinking is an essential component of human life. We are naturally interested in a systematic study of the activity of thinking. However, on account of its private and ephemeral character, we have come to distrust unaided reflections on our consciousness and to depend more on the study of the accompanying behaviours and the physiology of the brain, which are publicly observable. Recently the use of computational models has been added to these older tools for the study of the mind.

Indeed, with the rapid development and the increasingly broad application of computers, the traditional problem of mind and matter experiences a revival in the guise of a sharper formulation in terms of computers and computation. It is of both practical and theoretical significance to determine the extent to which thinking is computational. The ambitious general question is to ask whether all thinking is basically computational. Or, to use a familiar but ambiguous formulation: can computers think?

The central component of what I call *computabilism* is the position which answers this question affirmatively. When we try, however, to think about the reasons for and against this position, we begin to realize that we are inclined to conflate different questions by using certain natural but hard-to-justify presuppositions. In particular, most people who study the question today tend to assume, without pausing to ask for reasons, some form of *physicalism* or naturalism. More specifically, the common assumption is psychophysiological *parallelism* in the sense of a one-to-one correlation between mental and physiological phenomena. Consequently, computabilism for the mental is identified with computabilism for brain processes.

In this essay I shall first try to formulate some problems about the thesis of parallelism and its related conception of (the scope of) science. In any case, regardless of the truth or falsity of the thesis of parallelism, it is helpful to distinguish computabilism for the mental from that for the brain, because the reasons for and against them are of quite different types. In particular, we know at present more about the mind than about the brain.

Once we make such a distinction, we see that computabilism for the brain is related to the general question about the computational character of physical and biological processes. What is often known as *physicalism* includes both the thesis of psychophysiological parallelism and one of *biophysical* parallelism, which postulates some sort of reducibility of the biological to the physical. However, for all we know, even if computabilism for the physical is true, it does not follow necessarily that computabilism for the biological is also true.

We see, therefore, that the position of computabilism involves a number of different problems. I shall here consider some of these problems and shall formulate a few more definite subproblems. I shall present my thoughts under the following four headings: (1) Parallelism and the scope of science: can minds do more than brains? (2) Computabilism for the brain: is the brain like a computer? (3) Computabilism for the physical: is physics algorithmic? (4) Computabilism for the mental: is all thinking computational?

The content of consciousness is available to each of us through our inner experience and our disciplined introspection. There are certainly many conscious processes that are not at present known to be representable in physiological or physical or computational terms. The challenge or the frustrating task is to arrive at a reasonable conjecture as to the truth or falsity of physicalism and computabilism on the basis of what we know, taking into consideration our gross ignorance about the full powers of brains and computers.

PARALLELISM AND THE SCOPE OF SCIENCE: CAN MINDS DO MORE THAN BRAINS?

Traditionally there has been more interest in mental phenomena than in physical phenomena. But the experience of the last four centuries or so has taught us that we are better at attaining reliable knowledge about the physical than about the mental. As a result, science becomes associated with the systematic study of the physical world. For instance, the great popularity of neurobiology today assumes that the best way of really understanding the mental is by way of the physical. It is even assumed that, once we fully know the brain, we shall then know everything about the mind as well.

At the centre is the widely accepted assumption of psychophysiological parallelism: the belief that minds and brains are equivalent in the precise sense that there is a one-to-one correlation between one's mental states and brain states. Even though most scientists and philosophers today—Dennett, Edelman, Hofstadter, Minsky, Penrose, Searle, to name just a few—take parallelism for granted, there are recent thinkers such as Husserl, Bergson, Wittgenstein, and Gödel who all regard parallelism as a *prejudice* of our time. Of course a prejudice is not necessarily false. Rather it is just a strongly held belief not warranted by available evidence, its intensity being disproportionate to the solidity of evidence for it.

The localization to the brain as the site of mental activities is not obvious from everyday experience. For instance, the same Chinese character, *hsin* or *xin*, often translated mind–heart of body–mind, stands for both the mind and the heart, suggesting that mental activities take place in the heart. The Greek philosopher Empedocles thought that the mind was in the heart, taken as the material organ.

Certainly we are nowadays all aware of the existence of empirical evidence which indicates that the brain is intimately associated with one's mental activities. For example, if some parts of the brain are damaged or removed or disconnected, certain mental activities cease: arrest of brain development occasions imbecility; a blow on the head may result in unconsciousness, etc. We are therefore justified in concluding that there is some correlation, that thought is in some sense a *function* of the brain. But there are different kinds of function, many of which fall far short of one-to-one correlation. For example, the square root of a number has two values rather than one; for all we know, the function may be, in the words of William James (1960, p. 291), 'permissive' or 'transmissive' rather than 'productive'.

If we think of the spectacular mental feats of Leonardo or Leibniz or Schubert or Dostoevsky or Einstein, it is hard to imagine how we can possibly discern enough physical difference between their brains and certain physically similar brains to give an adequate account of the dramatic difference in their outputs.

Even in ordinary experience, it is difficult to exclude the possibility that our minds make more distinctions than our brains. For example, I have tried to reconstruct my discussions with Gödel in the 1970s with the help of certain rough notes taken at the time. I have been reminded of many things by these notes. My brain presumably contains in some sense certain traces that are not represented in the notes. But it is also possible that my mind remembers more than these traces in the brain.

Wittgenstein suggested a thought experiment that seems to give a more precise formulation of such a possibility. He imagined someone making jottings—as a text is being recited—that are necessary and sufficient for the person to reproduce the text. 'What I called jottings would not be a *rendering* of the text, not so to speak a translation with another symbolism. The text would not be *stored up* in the jottings. And why should it be stored up in our nervous system?' (*Zettel*, 612). In other words, reminders of all sorts, whether stored on paper or in the brain, may suffice to bring about a correct reproduction. It would be begging the question to say that this is possible precisely because the brain, unlike the paper, contains counterparts of everything in the mind.

Wittgenstein found it natural to suppose 'that there is no process in the brain correlated with associating or with thinking; so that it would be impossible to read off thought-processes from brain-processes', just as 'it is impossible to infer, from the properties or structure of the seed, those of the plant that comes out of it' (*Zettel*, 608). He considered recalling a man's name on seeing him again, and asks (*Zettel*, 610), 'Why should there not be a psychological regularity to which *no* physiological regularity corresponds?'

Indeed, he questioned, explicitly and in general terms, the ideas surrounding the 'prejudice in favour of psychophysical parallelism'. 'One of the most dangerous of ideas for a philosopher is, oddly enough, that we think with our heads or in our heads' (*Zettel*, 605). 'The idea of thinking as a process in the head, in a completely enclosed space, gives it something occult' (*Zettel*, 606). 'It is thus perfectly possible that certain psychological phenomena *cannot* be investigated physiologically, because physiologically nothing corresponds to them' (*Zettel*, 609). 'The prejudice in favour of psychophysical parallelism is a fruit of primitive interpretation of our concepts' (*Zettel*, 611).

Bergson believed (1946, p. 86) 'that between the consciousness and the organism there was a relation that no reasoning could have constructed a priori, a correspondence which was neither parallelism nor epiphenomenalism, nor anything resembling them'.

Gödel did not mention 'parallelism' in his discussion with me but considered instead the proposition: (1) There is no mind separate from matter. According to his belief, '(1) is a prejudice of our time, which will be disproved scientifically' (Wang 1974, p. 326). I shall not try to interpret the ambiguous statement (1). Since Gödel's specific conjecture of a disproof of (1) would also be a disproof of parallelism, I can reformulate his conjecture in terms of 'parallelism':

(A) Parallelism 'is a prejudice of our time which will be disproved scientifically—perhaps by the fact that there aren't enough nerve cells to perform the observable operations of the mind'.

In this context, *observable* operations undoubtedly include the operations observable by introspection, even though it is commonly assumed that scientific observations should be *publicly* observable. As we know, what is observable by introspection can often be communicated to others in such a way that they can also test the assertions by analogous introspective observations. Hence, there is no reason why introspective evidence should be excluded altogether. Indeed, discovering and understanding a proof or a hypothesis, for instance, all depend on introspection.

The conjecture that there are not enough nerve cells to perform the observable mental operations illustrates significantly a test of what we mean by a problem being *scientific*. Certainly the capacity of nerve cells is a natural and central topic of study for neuroscience. At the same time, the observable operations of the mind are also things that we are capable of knowing. These two beliefs or facts are undoubtedly the reason why the initial response of most of us is to agree with Gödel to the extent that the conjecture is indeed a *scientific* one, apart from the question whether it is plausible or not.

What is attractive about the conjecture is the apparent sharpness of its quantitative flavour. But the number of neurones in a brain is estimated at 10^{11} or 10^{12}, and there are many more synapses. We are not accustomed to dealing with the implications of such large numbers of units and our knowledge of their actual capacities—in particular, of the expected gap between their combinatorially possible and their actually realizable configurations—is very limited indeed.

Moreover, as we know, the mind increases its power by tools such as pencil, paper, computers, etc., by learning from others, and by using books and one's own writings as a kind of *external* memory. We have little knowledge of the extent to which the brain does the same sort of thing too. The situation with the brain may, therefore, be less clean than Gödel seems to think. Likewise, on the mind side we are at present far from possessing promising ideas of any kind to guide us in the quantitative determination of all *the observable operations of the mind*.

However that may be, it does seem to me that Gödel's conjecture suggests a possible connection between the 'purely philosophical' issue of psychophysiological parallelism and certain *scientific* considerations. Even for those who are unwilling to concede the scientific character of the conjecture, the proposal offers a challenge to them to produce some convincing reasons for denying its scientific status.

Usually the issues of parallelism and physicalism are discussed in the context of 'reductionism'. We often get caught in the frustrating task of distinguishing different senses of 'reduction'. One advantage of Gödel's conjecture (A) is that it bypasses this task, since, if and when it is seen to be true, we would have a strong conclusion, with a clear sense, that *no reduction* is possible. Establishing such a conclusion would cut through the ambiguity by ruling out even reducibility in the weakest acceptable sense; irreducibility would be proved independently of the ambiguity.

Even for those who believe that (A) is very unlikely to be true, the mere fact of our willingness to envisage the possibility of its being true helps to clarify our belief that the psychological and the physiological are qualitatively different, and throws doubt on the opinion that their equivalence can be seen by purely conceptual considerations.

One reason for the belief in parallelism is undoubtedly an inductive generalization from the surprising success of physics, not only in dealing with physical phenomena, but also in dealing with biological issues. Yet, given our experience of the striking distinction between the mental and the physical, it is by no means clear that, of the comprehensive assertion of parallelism, how typical a part of it our limited knowledge of correlation makes up.

At the same time, we are struck by the contrast between the maturity of our study of the physical world and the primitive state of our attempts to deal directly with mental phenomena. There is then a natural inclination to equate science with the science of the physical world. One is consequently inclined to believe that, if mind cannot be explained in terms of the science of the physical world, then it cannot be explained in *any* scientific terms at all. Given such a belief, one tends to regard the viewpoint of questioning parallelism as recommending a mystical direction or an obscurantist spirit.

Even though we have so far failed to study mental phenomena nearly as effectively as physical phenomena, this fact in itself is not a proof that parallelism is true or even irrefutable. We do not know that we shall not be able to study mental phenomena directly in an effective manner. At any rate, it seems clear that the truth or falsity of parallelism is to be decided on empirical evidence. At present, it is no more than an unjustified assumption of current science—rather than a probable conclusion consciously based on empirical evidence. Gödel's conjecture (A) is only one example of the kind of consideration that will help to clarify the status of the thesis of parallelism.

A more difficult problem than the insufficiency of evidence for the truth of parallelism is that of the fruitfulness of the belief that parallelism is not true. Given the fact that most current science assumes parallelism implicitly and most scientists are accustomed to working within the established framework, it is natural to find the questioning of parallelism little more than an idle intellectual exercise.

In order to render the denial of parallelism, or at least the suspension of belief in it, attractive to more people, it seems necessary either to find a conclusive refutation of it or to come up with serious and productive research programmes that contradict, or are independent of, the thesis of parallelism. There are indeed attempts to do so, but the outcome so far has not been unambiguously successful. Moreover, some of the programmes are linked to subjects and positions such as religion and mysticism, which are generally regarded as the opposite of science.

For example, James (1960, p. 279) connected his 'transmission theory' of cerebral action, a possible alternative to parallelism, to human immortality. In addition, after devoting twenty-five years to trying to make parts of physical research scientific, he acknowledged that, like Henry Sidgwick who had previously the same experience, he remained 'in the same identical state of doubt that he started with'. He then concluded (p. 310): 'at times I have been tempted to believe that the creator has eternally intended this department of nature to remain *baffling*, . . . my deeper belief is . . . that we must expect to make progress not by quarter-centuries, but by half-centuries or whole centuries'.

Einstein seems to have regarded telepathy as a physical phenomenon. After an extended discussion of the interpretation of quantum theory he chose to close it (1949, p. 683) 'with the reproduction of a brief conversation with an important theoretical physicist. He: "I am inclined to believe in telepathy." I: "This has probably more to do with physics than with psychology." He: "Yes"—.'

The fundamental obstacle in the way of studying mental phenomena scientifically is the fact that they are *subjective* but science is to be *objective*. That is why in the study of psychology we typically appeal to behaviours, to physiology, to computational models, etc. But it is clear that there is something special in our consciousness, in the subject as an agent, which, so far as we know, eludes such stable traces and representations of our consciousness.

Husserl spent much of his life trying to develop a science of the subjective, under the name of *phenomenology*. It has been very influential in philosophy but has, up to now, not developed into what we would accept as a science in the sense of being a stable and cumulative body of knowledge. There are a number of persuasive ideas in Husserl's discussions of the limitations of what he calls 'naturalism'. But his specific proposals as to how we can overcome these limitations and develop a new science are hard to disentangle. The liberation that he aims at is attractive (1970, p. 265): 'Phenomenology frees us from the old objectivistic ideal of the scientific system, the theoretical form of mathematical nature of science, and frees us accordingly of an ontology of the soul which could be analogous to physics.'

Gödel defends, in a manuscript from about 1962 (see Gödel 1994), the vision of a fruitful phenomenology by an analogy between it and physical science. The latter is, he says, a systematic and conscious extension of the child's experimenting with the objects of the external world and its own sensory and motor organs. Phenomenology should be an analogous extension of the child's coming to a better and better understanding of language and of the basic concepts on which it rests.

I have already mentioned Wittgenstein's qualms over parallelism. At the end of his *Philosophical investigations* (1953), he speaks of 'the confusion and barrenness of psychology' and blames it on *conceptual confusions*: 'The existence of the experimental method makes us think we have the means of solving the problems which trouble us; though problem and method pass one another by.' As we know, much of his later work may be said to be concerned with philosophical psychology or the philosophy of psychology. It is likely that he saw this work as helpful to the study of psychology. But it is not clear to me how much it has influenced the development of psychology so far.

It would be of interest to discuss the relevance of the ideas of Husserl and Wittgenstein to the possibility of a rewarding systematic study of our consciousness. A more specific topic is to consider the evidence and the reasons for and against psychophysiological parallelism and to ask whether parallelism is a scientific or a philosophical issue or both. As an example, Gödel's conjecture (A) provides us with a fairly definite focal point.

The approach I am adopting is to see the issues of parallelism and computabilism for the mental as the question whether minds are superior in the sense that there are things that minds can but bodies or computers cannot do. Consequently, we may choose any type of remarkable performance of the mind and challenge the parallelist or the computabilist to convince us that bodies or computers are also capable of such performance.

One natural choice is to consider mind's mathematical capabilities, which are comparatively definite and precise and, in particular, closely connected with computation. If we can give persuasive reasons for believing mind's superiority in mathematics the general belief in mind's superiority then follows as a consequence of the special case. Those who find the reasons insufficient may still try to establish mind's superiority by considering other types of mental phenomenon.

There are certain mathematical results, notably those of Gödel and Turing, that are directly relevant only to computabilism; they will be discussed in the final section. Other aspects of mind's mathematical powers pose problems for both parallelism and computabilism. Without giving details, I should like to indicate briefly some remarkable features of the development of our mathematical intuition that seem hard to duplicate by bodies or computers.

The human mind constantly develops, individually and collectively. As the mind develops, intuition develops too. Concepts are formed, and we arrive at forms and the formal by way of generalizing idealizations. Through the dialectic of the intuitive and the formal, of intuition and idealization, we are capable of extending our computation from small numbers to large ones, of making the *big jump* from the finite to the infinite, etc. In the process, we arrive at a surprising degree of objectivity and perspicuity. It is hard to imagine how brains or computers could be capable of accomplishing such amazing feats.

COMPUTABILISM FOR THE BRAIN: IS THE BRAIN LIKE A
COMPUTER?

It is 'very likely', Gödel asserted in 1972, without elaboration (Wang 1974,
p. 326), that:

(B) The brain functions basically like a digital computer. This is the thesis of
computabilism for brains. It is familiar in the study of artificial intelligence and
cognitive science, usually accompanied by a belief, often unstated, in
psychophysiological parallelism, contrary to Gödel's position. For instance,
Marvin Minsky affirms that '*Minds are simply what brains do*', and then says
(1988, pp. 287–8): 'There is not the slightest reason to doubt that brains are
anything other than machines with enormous numbers of parts that work in
perfect accord with physical laws.'

What do we mean by the thesis (B)? To explicate it calls for a lot of
idealization to eliminate certain obvious ambiguities. It envisages a computer
such that an idealized brain functions basically like it. A computer has certain
data stored inside it and certain openings to take in and put out data. It
operates on the inputs and the internal data to arrive at new states and
produce outputs from time to time.

The brain may be said to function basically like a computer M if there is some
sort of one-to-one correspondence between its 'significant' (in some suitably
chosen sense) components and states and those of the computer M. A simple
part of this somewhat vague condition is to require that corresponding inputs
produce corresponding outputs by them. More inclusively, corresponding
states are followed by corresponding states. Undoubtedly this rough
interpretation of (B) can be improved, but it gives us an idea of what is vaguely
intended by (B).

To believe (B) is just to believe that some such computer M exists, whether or
not we shall ever find it. A stronger belief would be that we shall probably find
it at some future time. Such a conjecture would seem to depend on existing and
projected empirical evidence and testing. The attempt to determine in practice
whether we can indeed find such a computer M may be seen as a large
research project. At the same time the assumption (B) or the belief that such a
computer exists serves also as a heuristic principle to guide our study of brains
and computers and their relations in more general ways than strategies for
specific tasks such as playing chess.

It seems clear that in considering (B) we become involved in both empirical
and conceptual investigations. Philosophy and science seem to interlock with
each other when we think about such problems. In any case, (B) seems to be a
meaningful proposition with some substantive content.

There is, however, one type of discussion in the literature which seems to deny that there is something like (B) which is a significant thesis with a real intent. For instance, Searle (1992, p. 200) discusses two theses or questions that appear to be similar to the thesis (B):

(B1) the brain is a digital computer (or, equivalently according to him, brain processes are computational);

(B2) the operations of the brain can be simulated on a digital computer.

Searle disposes of (B2) quickly, by asserting that it is obviously true, and concentrates on a discussion of (B1). His conclusion says (p. 225): 'The point is not that the claim "the brain is a digital computer" is false. Rather it does not get up to the level of falsehood. It does not have a clear sense.'

So far as I know, Searle does not consider (B). I am not sure how he would relate (B) to (B1) or (B2). Regardless of the question of a satisfactory formulation, I am convinced that (B) points to a meaningful and substantive question. It seems to me that those who see (B1) and (B2) as real and non-trivial questions would interpret (B1) and (B2) more or less as intending the same thing as my interpretation of (B). They would probably take (B1) to be but a less careful formulation of (B), and treat (B2) as equivalent to (B).

Searle's disposal of (B2) as trivially true seems to depend on his interpretation of the word 'simulate'. His argument is brief (Searle 1992, p. 200): (1) 'But given Church's thesis that anything that can be given a precise enough characterization as a set of steps can be simulated on a digital computer, it follows trivially that the question has a trivial answer.' This argument presupposes: (2) there is such a characterization of brain processes. His interpretation of Church's thesis is obscure or at least ambiguous. It is by no means obvious that (2) is true in the sense of there being a characterization satisfying the requirement of (1).

Searle seems to take 'simulate' to mean 'approximate'. But there are approximations and approximations. Those who take (B2) to be a serious thesis undoubtedly have in mind a certain *completeness* of simulations, so that the desired simulation has to be exact or at least an approximation with a suitably high degree of accuracy. I am inclined to think that, even though Searle's assertions may be true of some interpretation of (B2), they are not true of the interpretation of it intended by those who propose (B2) as a thesis or a conjecture.

Searle's extended discussion of (B1) emphasizes that 'syntax and symbols are not defined in terms of physics' and that they are 'observer relative' (p. 225). His central argument for questioning the meaningfulness of (B1) appears to me to be the following contrast (p. 214): When I use a computer, there is no homunculus problem; if the brain were a computer, who would be the user? I take his conclusion to be: since a brain is its own user, it cannot be a computer. For those who, like Gödel, distinguish the mind from the brain, there is a ready answer: The mind is the user of the brain functioning as a computer.

Searle appears to be among those who do not make such a distinction but attribute all mental categories (such as 'intentionality') to the brain (as well as the mind). His observations about the homunculus and about syntax and symbols seem to imply that, for him, the brain is not (just) a physical object. He seems to say that neither a computer nor indeed any physical object could embody 'the concrete biological reality of intrinsic intentionality' of the brain (p. 224). However, the situation is by no means clear to me. For example, we do not know enough about computers to exclude the possibility that there is a reasonable sense in which some computer may serve as its own user, know its own program completely, etc.

As far as I know, those who assert (B) have not offered any convincing arguments to support their conclusion. The basic idea appears to me to be the belief that brains cannot but be computers: being physical organs, what else can they be? The only serious problems, as Minsky says, 'come from our having had so little experience with machines of such complexity'. More explicitly, Churchland and Sejnowski (1992) seem to take (B) to be true by definition (p. 61 of their book): 'Notice in particular that once we understand more about what sort of computers *nervous systems* are, and how they do whatever it is they do, we shall have an enlarged and deeper understanding of what it is to compute and represent.'

Even though Gödel offers no direct argument to support (B), he seems to suggest, all too briefly, the idea that a brain, being a physical organ, is capable of only a finite number of states and, therefore, can perform only mechanical procedures (Wang 1974, pp. 325–6). Since his argument involves the nature of physical laws and Turing's definition of mechanical procedures, I shall examine it in the final section.

One possible line of trying to disprove (B) is to ask whether physics is algorithmic and, if not, whether its noncomputable component plays an essential part in the functioning of the brain. One example of pursuing this approach is the programme of Penrose elaborated in *The Emperor's new mind* (1989) and in a new book in preparation. He speculates that 'a yet-to-be discovered theory of quantum gravity' may turn out to be nonalgorithmic and that there is then some possibility for a nonalgorithmic ingredient in some type of brain-action (Penrose 1989, pp. 438–9). As far as we know now, it is not excluded that Penrose's specific programme will not work or even that the general *type* of disproval of (B) along such a line can be shown to be impossible. At the same time, we also do not know enough to say that such attempts cannot be fruitful.

The brain is certainly not an ordinary physical object, and it is by no means clear that, even if all physical processes are algorithmic, brain processes must also be so. In terms of the laws of physics, the Darwinian laws of selection, in evolution and in individual development, interpreted in one way or another, are certainly not consequences of them. There is, therefore, the additional question whether the laws of selection are also algorithmic, even if we assume that the laws of physics are algorithmic and that the laws of selection are sufficient to account for brain processes. A somewhat different question is whether the brain, in its processes, is essentially different from (other) matter.

Another approach toward a disproval of (B) is to try to show that, regardless of whether physics is algorithmic, biology is not. One example along this direction is Edelmans's work, as summarized in his 1992 book. According to him, biology is basically different from physics. In particular, he emphasizes that, in contrast to physics, biology as we know it is specific, it is *historical*: 'Evolution is based on a *particular* historical sequence of natural selection from populations of variant organisms' (Edelman 1992, p. 213). A pervasive theme of the book is that 'the brain is not a computer and the world is not so unequivocally specified that it could act as a set of instructions' (p. 189).

So far as I am able to determine, Edelman does not offer any precise argument to refute (B). Instead, he points out that both events in the environment and the generator of diversity—mutations, alterations of neural wiring, or synaptic changes—in the brain, as a selective system, are *unpredictable* (p. 190). It would be of interest to try to see whether we can formulate these unpredictable components precisely enough to be able to extract from the formulation a proof that the brain cannot indeed be a computer.

In connection with Darwinism and the brain, there are three large questions that most practising scientists do not ask, undoubtedly because they are remote from their immediate work: (1) whether Darwinism suffices to account for brain processes; (2) whether the laws of selection—evolution and individual development—are algorithmic; (3) whether the brain, in its processes, is essentially different from (other) matter.

When we do face these questions, we see that there are different opinions about them. For instance, Ashby, like Edelman, assumes (1) and sees the brain as an organ developed through selection, but chooses different answers to (2) and (3). He proposes to idealize the brain as a machine and assumes that the nervous system, and the living organism in general, in its nature and processes, is not essentially different from other matter (Ashby 1960, pp. 8 and 29–30).

Even though (1), the adequacy of Darwinism, is widely accepted, there are notable dissenting opinions. For instance, Wittgenstein, on one occasion while admiring the immense variety of living things in a zoological garden, said

(Rhees 1984, p. 160), 'I have always thought that Darwin was wrong: his theory doesn't account for this variety of species. It hasn't the necessary multiplicity.'

Gödel also questions (1), and seems to affirm (2), the algorithmic character of the laws of selection. On 13 October 1971 Gödel expressed to me his position on Darwinism more or less in the following words:

I don't think the brain came in the Darwinian manner. In fact, it is disprovable. Simple mechanism can't yield the brain. I think the basic elements of the universe are simple. Life force is a primitive element of the universe and it obeys certain laws of action. These laws are not simple and not mechanical.

In a later session, Gödel said something like this:

Darwinism does not envisage holistic laws but proceeds in terms of simple machines with few particles. The complexity of living bodies has to be present either in the material or in the laws. In particular, the materials forming the organs, if they are governed by mechanical laws, have to be of the same order of complexity as the living body.

In the version prepared carefully by himself for publication, Gödel omitted the mention of 'life force' and elaborated his idea about the human body (including the brain) as a generalization of his conjectured disproval of the belief that minds are equivalent to brains. He chose to express his views in the third person (Wang 1974, p. 326):

More generally, Gödel believes that mechanism in biology is a prejudice of our time which will be disproved. In this case, one disproval, in Gödel's opinion, will consist in a mathematical theorem to the effect that the formation within geological times of a human body by the laws of physics (or any other laws of a similar nature), starting from a random distribution of the elementary particles and the field, is as unlikely as the separation by chance of the atmosphere into its components.

Gödel presumably saw the power of the mind as a distinguished part of life force that is higher than and not reducible to the exercise of the life force by the human body or the brain, a process that is the subject-matter of biology. The phrase 'more generally' seems to say that not only the mental part is irreducible to the bodily part of the life force, but also, more generally, the bodily life force cannot be accounted by Darwinian evolution alone.

As I said before, Gödel believed it very likely that (B) the brain functions basically like a digital computer. This would seem to suggest that it was for him in principle possible that the brain or the human body could be formed in a mechanical manner. His conjectured disproval of 'mechanism in biology' would seem to be directed against a specific manner of mechanical formation that he associated with Darwinism. He seems to be saying that, even though

(B) may be true, the brain is so complex that it is most unlikely that it has been formed in the manner determined by Darwinism: that the laws of Darwinism, as he understood them, are of such a type that a computer as complex as the brain is highly unlikely to be formable in geological time.

We know more about the physical than about the brain. By treating the brain as a physical object, we reduce or at least relate the thesis of computabilism for the brain to that for the physical, on which we have something more definite to say. In this way we bring about a separation of the physical aspect of the brain from its (connection with the) mental aspect (of a person). And this conforms to the wisdom of dividing up the difficulties, so that each part can be overcome more easily. Those who uphold psychophysical parallelism can view the distinction between the mind and the brain as a distinction between two aspects of the brain that are being considered separately for the purpose of making more effective use of what we know now.

In view of these considerations, I am in favour of adopting the strategy, at least under the present circumstance of our ignorance, of concentrating our attention on the issue of computabilism for the physical and that for the mental. We have, I believe, more information about these two issues than on the issues of psychophysiological parallelism and of computabilism for the brain, which we have considered in this and the preceding sections.

COMPUTABILISM FOR THE PHYSICAL: IS PHYSICS ALGORITHMIC?

We all recognize a distinction between physical processes as they are and our knowledge of them. We distinguish between the real, the known, and the knowable. Our knowledge of the physical processes depends on our observations, as a more or less direct contact between us and the physical world. We arrive at physical laws by reflecting on our observations, which in turn are compared with the observable consequences of the physical laws, as a way to verify our support and to falsify or throw doubt on them. Therefore, in considering the physical, we have to take into consideration the interplay of three factors: the physical processes themselves, our physical laws, and our observations. In order to interpret the thesis of computabilism for the physical, we have to pay attention to this interplay.

When we measure length or time or weight or temperature, etc., we use numbers to express our results. We are not able to discern arbitrarily small differences, even when we use instruments like microscopes, telescopes, etc. I believe that the near edge of my desk has a real length. But in measuring it carefully, the number I get is accurate only to a certain degree: I would say, for instance, it is significant to three digits. When I measure it again and again, I am likely to get different results that agree only up to the first three digits. Let me express this gross experience in a principle:

(C) Our observations have a finite limit of precision.

The physical laws are usually expressed in the precise language of mathematics. They are not and cannot be determined completely by observations, yet they have to agree quite well with our observations to be accepted. This means in part that their *observable consequences*, say the position of a physical object at a later time as predicted by the laws on the basis of its position at an earlier time, agree pretty well with the position observed at the later time. Even though the physical laws may be precise, their applications to the real world are not entirely precise, because the earlier and the later positions, as observed and numbered by us, have some finite limit of precision. The value predicted by the laws can agree with our observed value only within some limit, and it may or may not agree completely with the value of the object's *real* position. This situation is very probably what Gödel had in mind when he said to me in 1971 that the following assumption is 'practically certain' (Wang 1974, p. 326):

(D) The physical laws, in their observable consequences, have a finite limit of precision.

As we know, computable (real) numbers constitute only a 'small' part (in terms of Cantor's definition of the size of infinite sets) of the real numbers envisaged in classical mathematics; and computable functions (of natural numbers or of real numbers or of functions of different types) are only a very special, though distinctly important, case of all mathematical functions. Yet, at least when mathematics is applied in physics, computable numbers and functions occupy a distinguished place—in view of the fact that any noncomputable number or function can be approximated by a computable one to any desired degree of precision.

For instance, if we think of an arbitrary real number as an integer followed by an infinite decimal, then, for every n, there are many computable real numbers that agree with it up to the first n decimal digits. Equivalently, for any function or sequence of natural numbers and an arbitrary number n, there are many computable functions that agree with it up to the first n terms. Similarly, with functions (say) from real numbers to real numbers, we have certain natural definitions of computable functions and can approximate each arbitrary function by a computable one to any desired degree of precision. For convenient reference, we may treat this observation as a principle:

(E) An arbitrary real number or function can be approximated arbitrarily closely by a computable one.

Since physical laws are generally formulated in terms of numbers and functions, the principles (C), (D), and (E) suggest the conjecture that the use of noncomputable (real) numbers and functions is a matter of convenience and that there is some sense in which we ultimately go only from the computable to the computable and from the finite to the finite. Indeed, this familiar idea has often been proposed and elaborated in the literature. My impression is, however, that it is hard to work out this idea to arrive at a clear and convincing position.

Physical theories and laws are our attempts to represent approximately the actual physical processes by means of our observations and our reason. In order to consider the question whether physical processes are algorithmic, we have to study first the question whether physical laws are algorithmic, since, in order to find out some property about the real world, we have to begin with what we know. This means that we have to ask first whether our current physical theories are algorithmic.

The best sense that can be made of this question is, I believe, to ask whether the predictions of the theory under consideration preserve computability. That is to say, to ask whether the predictions are always computable when the input data are computable.

The principles (D) and (E) do not exclude the possibility that some physical theories do not preserve computability. We have to examine them individually to determine whether this is the case or not. There is some work along this direction in the literature. For instance, it is proved in Theorem 9 of Pour-El and Richards (1983) that there exists a solution of (a particular form of) the wave equation such that the output at time 1 is not computable in the input at time 0. The solution is, however, physically unrealistic. (Apparently the phenomenon can occur only with solutions that are not twice differentiable and therefore physically inappropriate.)

Even though I have seen no systematic consideration of the issue of computabilism for the physical, I am under the impression that the consensus of those who have thought about the issue is that the physics of today is algorithmic. For example, Penrose concludes from his study that classical theory, including the theory of relativity, is algorithmic (Penrose 1989, p. 216), and Wayne C. Myrvold told me in correspondence that 'some of the more obvious ways one might attempt to produce noncomputable predictions from computable initial conditions in quantum mechanics are bound to fail'.

The combination of the principles (D) and (E) appears to imply that, relative to what is observable, it is always possible to find algorithmic physical theories. The finite precision of observation seems to impose a veil or a sieve between physical theories and physical reality that conceals or excludes from physical theories whatever noncomputable elements may in fact exist in physical reality. Since principle (D) says that we are not sure of the absolute faithfulness of any physical theories to physical reality that transcends all finite limits of precision, principle (E) seems to entail that there are always computable theories that agree with our observations of physical reality, whether or not physical reality in fact contains noncomputable elements.

Something like this may be what Gödel had in mind when he appeared to argue that, on account of (D), if parallelism is true, then mind can perform only mechanical procedures (Wang 1974, pp. 325–6). Gödel seems to imply also that, on account of (D), it is 'practically certain' that the physical laws are and will continue to be algorithmic.

Regardless of whether my interpretation of Gödel's condensed observations is correct, I have found the general line of thought puzzling. Even though physical theories are ultimately to be checked against observations, they go far beyond our actual observations and take into consideration only a suitably selected representative fragment of all our observations. We have in fact other requirements, such as simplicity and perspicuity and conformity to general principles of logic and other deeply rooted beliefs, which go beyond the verifiability condition of the agreement of their consequences with our observations of limited precision. What is knowable, even though its consequences have to agree with observations, goes far beyond our observations, which constitute only an elementary component of our knowledge and thinking.

Recently Roger Penrose commented, in correspondence, on my report of Gödel's views, in words that seem to communicate more clearly what I am trying to say here.

One hardly ever tests a theory in terms of the detailed evolution from initial data, for example. Experiments often test indirectly, and may involve the general consistency of the ideas of the theory as a whole. A theory might have noncomputability as one of its by-products, whereas the general principles leading to this noncomputability might themselves be tested in other ways.

For instance, it is not clear to me how the principle (D) applies to properties that have only yes and no answers or to questions that depend only on less-than-precise observations. There may be physical connections that magnify small differences and reveal larger observable differences. We may be able to find reasons to suspect that physical reality, as it is, is or is not representable, to arbitrary degrees of precision, by an algorithmic theory.

A basic problem in considering whether physical processes are algorithmic is the fact that computability deals with infinite sequences but our experience is fundamentally finite. One question has to do with conspicuous events like earthquakes, eclipses, hurricanes, etc. For example, consider earthquakes clearly above a fixed lower bound and ask the question whether, for each n, beginning one thousand years before now and continuing into the future, such an earthquake begins on the nth day, leaving out cases that begin too close to the borderline between one day and the next. For all we know, these numbers may form a noncomputable sequence and a physical theory may have such a prediction.

It seems to me possible, despite principle (D), to conceive of both such a theory and increasingly more convincing confirmation of such a prediction: we can, I think, contrive examples of theories and their consequences in such a way that, despite the limited precision of our observations, the situation envisaged by Penrose is possible. Others can probably come up with more persuasive examples than mine about the earthquakes.

Consider another question: is it possible to find or construct a material box such that, for a certain computable sequence of inputs, its outputs form a noncomputable sequence? There are two questions: whether this is possible and whether, if we have such a box, we can know that it is indeed such a box. I am not able to see that our present knowledge of what physics can be, as Gödel suggests, excludes the possibility of finding or constructing such a material box, not only as a matter of fact but also observably so.

As I see it, the principle (D) points only to a limitation of our ability to observe, in contrast to that of thinking in general. Since observations are only one type, and an 'elementary' one, of our mental activities, it is natural to expect the possibility that the intrinsic limitations on them can be transcended with the help of other ways of thinking about physical reality, on the ground of our general experience that our reason has powers in certain aspects beyond our senses.

The purpose of physical theories is to capture as fully as we can our experience of what is going on in physical reality. But this experience is not exhausted by physical theories, because it also produces in us certain broad beliefs about a number of general features of physical reality and of our thought about it. For instance, we believe that the laws of logic are applicable to them and that mathematics is helpful in the development of physical theories.

In the context of discussing the Einstein–Podolski–Rosen paradox, Einstein makes explicit (1949, p. 682) two general beliefs, commonly referred to as 'realism' and 'locality', which are more intimately connected with the actual pursuit of physics than our belief in the applicability of logic and mathematics. As a result of the introduction and confirmation of *Bell's inequality* in recent years, it is now generally accepted that quantum theory is indeed a non-local theory. Bell and a number of other physicists see this as a dilemma that calls for a revision of quantum theory. This illustrates the fact that, even when the observable consequences confirm a theory, we may still question it on the ground of certain general principles—in this case the principle of locality. One question relevant here is whether the observed non-local correlation between two quantum systems might provide a possible course for avoiding the restriction imposed by the combination of principles (D) and (E).

COMPUTABILISM FOR THE MENTAL: IS ALL THINKING COMPUTATIONAL?

A familiar formulation of the question is: Can computers think? Obviously this formulation is highly ambiguous. Since thinking includes so many things, it seems necessary to single out certain representative types of thinking in order to give the question some definite sense or to deal, initially at least, with some definite aspects of the question. There are a number of different proposals to meet this need. One famous example is Turing's 'imitation game', which asks whether computers can carry on an intelligent conversation with human beings or with one another. Another example is to ask whether computers can do as much mathematics as human beings.

If we find some mathematical tasks, such as proving theorems or finding axioms or formulating conjectures, which human beings can do but no computer can, we can then conclude that minds are in some way superior to computers in doing mathematics. Moreover, we can then conclude generally that minds are not equivalent to computers in their capacities. In that case we would also have a definite sense according to which the original question is answered in the negative: computers cannot think, in the sense that they cannot think as well as human beings in all directions.

Of the issues discussed in this essay, computabilism for the mental is the central one and it has been considered most extensively for many decades. In particular, both Gödel and Turing, who have each contributed so much to the clarification of the nature of computation, gave much thought to the question and came out in favour of opposite answers to it. Gödel wished to deny and Turing wished to affirm that all thinking is computational. But neither of them claimed to have a conclusive proof for his position.

Both of them considered the implications of Gödel's and related results on the limitation of formal systems and the equivalent notion of computational procedures, as explicated by Turing and several other logicians. In addition, Gödel discussed Turing's arguments for the adequacy of his characterization of human computation—to make a distinction between mechanical procedures and mental procedures.

For our present purpose, we may confine our attention to *the Gödel theorem which says that in no consistent formal system can the proposition expressing its own consistency be proved.* Yet by knowing Gödel's theorem, we seem to be able to prove the proposition. Hence, we seem to be able to do better than any given computer. The problem is that, contrary to our initial impression, this line of thought does not yield what we are after—at least not without more ado.

In his Gibbs lecture, 'Some basic theorems on the foundations of mathematics and their philosophical implications' (Gödel 1994), delivered to the American Mathematical Society on 26 December 1951, Gödel observed that the theorem does not preclude the possibility that there should be a rule or algorithm equivalent to our mathematical intuition: 'However, if such a rule exists,' that the propositions it produces 'are all true could at most be known with empirical certainty.' In this context Gödel adds a footnote which suggests that his views against parallelism and computabilism were much more hesitant than in the 1970s:

For example, it is conceivable (although far outside the limits of present-day science) that brain physiology would advance so far that it would be known with empirical certainty: (1) that the brain suffices for all mental phenomena and is a machine in the sense of Turing; (2) that such and such is the precise material structure and physiological functioning of the brain which performs mathematical thinking.

In 1972 Gödel summarized this line of thought in the following words (Wang 1974, p. 324):

The human mind is incapable of formulating (or mechanizing) all its mathematical intuitions. I.e.: if it has succeeded in formulating some of them, this very fact yields new intuitive knowledge, e.g. the consistency of this formalism. This fact may be called the 'incompletability' of mathematics. On the other hand, on the basis of what has been proved so far, it remains possible that there may exist (and even be empirically discoverable) a theorem-proving machine which in fact *is* equivalent to mathematical intuition, but cannot be *proved* to be so, nor even be proved to yield only *correct* theorems of finitary number theory.

In other words, there is at least no provably adequate and correct algorithm that is equivalent to mind in its mathematical capacity. Thus, if there were such an algorithm, since we know that all its theorems are true, we also know that it is consistent, contrary to the assumption that it contains all the true propositions knowable by our mathematical intuition.

Even though this argument is generally accepted, it does involve a subtle element involving mind's constant development. For instance, we have a fairly standard formal system, known as 'the first-order theory', of the

Dedekind–Peano arithmetic. We are convinced of the truth of its axioms and rules, but it is a major undertaking to carry out an explicit proof of its consistency. Indeed, it is the main point of Gödel's theorem that such a proof cannot be executed in the formal system itself. The difference between a given formal system and our mathematical intuition is exactly the fact that, unlike a formal system, our intuition, once it understands something fully, has the natural tendency of going beyond it. Therefore, we may accept, as our starting-point, the following fact:

(F) There exists no provably correct algorithm that is provably equivalent to mind's mathematical intuition.

In order to prove mind's superiority, we have to strengthen (F) by deleting the two occurrences of the requirements *provably*. In other words, the problem is whether there may exist an algorithm that is *in fact*, though not provably, correct and equivalent to mathematical intuition.

Suppose I set up a formal system and claim that I perceive its axioms and rules to be true and believe that they contain all *our* mathematics. Since this contradicts (F), there must be something wrong with my claim. In the first place, if the system is in fact inconsistent, my belief is wrong. In the second place, if it is consistent, then I am wrong because the statement of consistency is part of my belief but not part of the formal system. The important point is rather that I cannot know an adequate formal system in the sense of perceiving its axioms and rules to be true: by definition, a formal system can be given by a finite number of axioms and rules, all within the range of mathematical intuition, so that we should be able to perceive them to be true one by one. We are, therefore, not able to set up a formal system which we know to be correct *and* which is in fact equivalent to mathematical intuition.

The only remaining possibility for mathematical intuition to be algorithmic is the existence of an unknown and unknowable algorithm which is in fact equivalent to mathematical intuition and which we use unconsciously. Even though we do not know enough to exclude such a possibility, it is hard to see how one can find specific reasons to support the belief in the existence of such an algorithm or supply plausible suggestions as to how such an algorithm could have formed in the development of the human species.

There are records to show that Turing began to think about computer intelligence and computer learning, as well as the effect of Gödel's theorem on the issue, at least as early as 1947. He repeatedly denied the conclusiveness of the theorem as a refutation of computabilism and suggested the idea of permitting the computer to give occasional wrong answers (Turing 1950, pp. 2109–10; 1959, p. 129; 1969, p. 4; 1986, p. 124).

In order to consider Turing's ideas about computers, which are, like human beings, capable of making mistakes as well as new and very interesting statements, we should, as he suggested, look at how mathematicians proceed to prove new theorems. For instance, for over 350 years, various proofs of Fermat's conjecture have been proposed and rejected. In 1992 a new proposal was made public and is being taken seriously in the mathematical community. We expect that within a year or two this proof will either be shown to be mistaken or be generally accepted. When we reflect on what is involved in the process, we realize that, if there is indeed an algorithm that is equivalent to the way in which mathematicians go about doing mathematics, it must be very complex indeed.

For Turing's purpose, it is sufficient if we can find a computer that performs as well as a pretty good mathematician. Yet, assuming that such a computer can be found, it is, as I have said above, unlike a human being in that its consistency is seen by us but goes beyond its own capacity to prove theorems. Turing probably had in mind the designing of a computer whose operations we do not fully understand. In other words, even though a theorem-proving computer uses, by definition, only finitely many axioms and rules, we may fail to perceive all of them as true. We may believe that they are probably true, so that the computer is probably consistent. It seems that, in such a case, its inability to prove its own consistency is not much different from our use of many formal systems, such as those for set theory, that we do not know to be consistent.

Even though we do not know with certainty even that the familiar axioms of set theory are consistent, we are interested in the quest for stronger and stronger axioms. The relevant point here is that our mathematical intuition is open-ended in its development, even when we are dealing with highly coherent and plausible beliefs that are not known with certainty. Consequently, if we believe the axioms of a strong formal system of set theory to be true, we also believe that the system is consistent. This is a feature not shared by the provable propositions in a formal system.

One familiar line of thought against computabilism is to try to use Gödel's theorem to distinguish minds from computers by showing that mind has enough self-knowledge to prove its own consistency. I have considered the ramifications of this idea without finding any conclusive answer (Wang 1974, pp. 317–21), and Gödel commented on my discussions by suggesting the programme of trying to distinguish legitimate from illegitimate uses of self-reflexivity (Wang 1974, p. 328, note 14). The idea is to understand the self-reflexive use of the concept of human proof in the proposition 'all human proofs form a consistent whole', so as to see that it is itself humanly provable. In other words, we hope to be able to see that, unlike the proofs of a computer, the human proofs include a proof of the proposition that the set of all human proofs is consistent. There are also problems about the self-knowledge of computers. For instance, Seymour Papert once told me that Gödel had asked, at a meeting in June 1972: Is there anything paradoxical in the idea of a machine that knows its program completely?

Gödel's theorem invites an extension of any given formal system by exhibiting true propositions that are not provable in it. If we begin with a familiar formal system, and some specific way of extending it, we may repeatedly make extensions, not only from one system to the next, but also by taking the union to merge all previous extensions at limit ordinals. In this way, by using infinite ordinal numbers, we are able to get richer and richer systems.

In his doctoral dissertation at Princeton, Turing (1939) explored this idea to study the question whether we may find a sequence of such systems that, taken together, would be complete. His idea was to confine the non-mechanical steps entirely to verifications that certain relations between positive integers do define ordinal numbers. What is involved is a vaguely defined procedure of defining recursive well-orderings of integers that represent larger and larger ordinals. What Turing said about intuition in this context (1939, pp. 208–9) seems to suggest a belief that we are not like computers.

Two of the main topics of the discussions between Gödel and me from 1971 to 1972 are (1) Turing's definition of mechanical procedures and (2) mathematical arguments on minds and machines, starting with a draft of two sections of my 1974 book (II3 on pp. 81–99 and X6 on pp. 315–24). On the one hand, Gödel regards Turing's definition as 'precise and unquestionably adequate': 'We had not perceived the sharp concept of mechanical procedures sharply before Turing, who brought us to the right perspective' (Wang 1974, p. 85). On the other hand, he took Turing's argument as containing also a fallacious proof 'for the equivalence of mind and machines' (p. 325). For related statements by Gödel, compare Davis (1965, pp. 71–3) and Gödel (1990, p. 306).

At the first of the scheduled sessions on 13 October 1971, Gödel directed his comments to my elaboration of what I took to be Turing's argument (as given in Davis (1965, p. 136), under the label 'the principle of finiteness' (later published in Wang 1974, pp. 92–3). The crucial point is that both Turing and I implicitly assumed the equivalence of minds and brains to argue for the thesis:

(G) A finite mind is capable only of a finite number of distinguishable states.

Gödel criticized the argument more or less in the following words (compare Wang 1974, p. 102, note 30):

Even if the finite brain cannot store an infinite amount of information, the spirit may be able to. The brain is a computing machine [situated in the special manner of being] connected with a spirit. If the brain is taken as physical and as a digital computer, from quantum mechanics there are then only a finite number of states. Only by connecting it to a spirit might it work in some other way. The thesis of finiteness presupposes: (1) spirit is matter, and (2) either physics is finitary or the brain is a computing machine with neurons as components. 'The spirit and the brain are not the same' is a weaker presupposition than 'the spirit and the brain are the same'.

Gödel then mentioned that he had a typed page on this question, which was forthcoming in *Dialectica* (undoubtedly the note 'A philosophical error in Turing's work' on p. 306 of Gödel 1990). Early in 1972 Gödel finally gave me a copy of an apparently revised version of the page, which was published in my 1974 book (pp. 325–6).

Both versions emphasize 'the fact that *mind, in its use, is not static, but constantly developing*', and aim at the same conclusion: 'although at each stage of the mind's development the number of its possible states is finite, there is no reason why this number should not converge to infinity in the course of its development'.

In both versions Gödel appeals to our experience of forming stronger and stronger axioms of infinity in set theory and of defining computable (or recursive) well-orderings of integers that represent larger and larger ordinal numbers. The point is undoubtedly to illustrate, by using the concept of set and the concept of ordinal number as examples, the fact that we do understand concepts (or abstract terms) more and more precisely as we go on using them.

In the later version, Gödel added two examples to illustrate the fact that more and more concepts (or abstract terms) enter the sphere of understanding. First: 'The iterative concept of set became clear only in the past few decades.' Second: 'Several more primitive ideas now appear on the horizon, e.g. the self-reflexive concept of proper class.' Gödel had in mind, I believe, his distinction between sets and concepts, according to which there are concepts whose ranges are not sets but proper classes. Since, according to Gödel and our intuitive belief, the concept of concept applies to itself, the proper class which is its range belongs to itself. That is the sense in which the concept of proper class is 'self-reflexive'.

Gödel's line of thought seems to move from conjecture to conjecture. (a) Mental procedures can go further than mechanical procedures. (b) The number of mind's states may converge to infinity in the course of its development. (c) 'Now there may exist systematic methods of accelerating, specializing, and uniquely developing this development, e.g. by asking the right questions on the basis of mechanical procedures.'

Gödel appeared to view (c) as one of the possible ways to prove (b). It is, however, not easy to envisage what a precise characterization of a systematic, but not mechanical, method, say of introducing larger ordinal numbers or stronger axioms of infinity, would look like. Gödel did recognize this difficulty.

Indeed, after stating conjecture (c), he continues: 'But it must be admitted that the precise definition of a procedure of this kind would require a substantial deepening of our understanding of the basic operations of the mind.'

I do not have a clear understanding of the meaning of (b) and its relation to (a). Gödel seemed to view (b) as both a more specific and a stronger assertion than (a). When we think about our mental states, we are struck by the feeling that they and the succession of them from one state to the next are not so precise as those of Turing machines or computers generally. Moreover, we develop over time, both individually and collectively, so that, for instance, what appeared to be complex becomes simple and we understand things that we did not understand before. Here again we feel that the process of development is somewhat indefinite and not mechanical. Yet we do not see how we can capture these vaguely felt differences in sufficiently explicit formulations to secure some rigorous proof that we can indeed do more than computers in certain specific aspects. Gödel's choice of conjecture (b) gives the impression of providing us with an exact perspective to clarify the differences, since the distinction between the finite and the infinite is one of the clearest differences we know, especially from our experience in mathematics.

But the relation between the contrast of the finite with the infinite and that of the mechanical with non-mechanical is not a simple one. Computers, each with only a fixed finite number of machine states, can in principle typically add and multiply any of the infinitely many numbers. It is not necessary for a mind or a computer to be in distinct states to deal with distinct numbers. Gödel's notion of 'the number of mind's states converging to infinity' is, I think, a complicated requirement, since there are states of different degrees of complexity. We seem to need a criterion to determine what sort of thing constitutes a state, in order to be able to count the number of states in any situation. For instance, we may try to specify a measure of simplicity such that what we think of as possible states of computers are all simple according to the measure. It is not clear to me what a natural and adequate measure might be, except perhaps the condition that the state be physically realizable.

Assume for the moment that some such criterion is given. How do we go about determining whether or not the number of mind's states converge to infinity? It would be hard to break up mental states into such simple states. An easier approach might be this: select certain things that minds can do and show that they require more and more simple states in the agreed sense. Ideally, of course, we would have a proof of conjecture (a) and of a stronger conclusion than conjecture (b), if we could find something that minds can do but cannot be done by using no matter how many *simple* states. If, however, we do not have such a strong result but have proved (b) only in terms of simple states (in the agreed sense), I am not sure that we would thereby also have a proof of conjecture (a).

Suppose we have found a proof of (b). Converging to infinity in this case means just that, for every n, there is some stage in the development of a mind such that the number of its states is greater than n. Since the states are, by hypothesis, of the kind that is appropriate to computers, there is, for each stage in the mind's development, some computer that has the same states as the mind at that stage. It remains possible that the different stages of the mind's development are related in a computable manner, so that there is a sort of supercomputer that modifies itself in such a way that, at each stage of the mind's development, the supercomputer functions like the computer that has the same states as the mind at that stage. Hence, it seems to me that the crucial issue is not whether the number of mind's states converges to infinity, but rather whether it develops in a computable manner.

Before trying to formulate a few definite problems, I should like to mention two difficulties that I have with regard to Gödel's reasoning. In the first place, he seems to say that once we accept psychophysical parallelism, computabilism for the mental is true. My own feeling is that we have no decisive evidence for this inference. It is possible that psychophysical parallelism is false but psychophysiological parallelism is true. Even if psychophysical parallelism is true, computabilism may be false for both the physical and the mental.

In the second place Gödel seems to say that once we accept Turing's argument, we have to accept the equivalence of minds and computers too. This is by no means clear to me. For one thing, Turing himself seems not to have made such a direct inference himself. Turing did once say: 'A man provided with paper, pencil, and rubber, and subject to discipline, is in effect a universal [Turing] machine' (Turing 1969, p. 9). But the statement is ambiguous and, in any case, Turing did not relate this observation to his argument about the limitation to finitely many states of mind.

Even though I am not able to understand Gödel's reasoning fully, it seems to suggest a number of fairly definite propositions and problems about their relations. We have: (A) psychophysical parallelism; (A′) psychophysiological parallelism; (B) computabilism for the brain; (D) finite precision of the observable consequences of physical laws; (G) a finite mind is capable only of a finite number of distinguishable states; (G′) a brain is capable only of finitely many distinguishable states; (H) mental procedures cannot carry any further than mechanical procedures; (I) computabilism for the mental; (J) computabilism for the physical.

I am identifying physicalism with (A), which is equivalent to (A′) plus a biophysical parallelism. The main problems about computabilism are (I) and its subproblems (B) and (J). Obviously (A) and (B) together imply (I); (A) and (G') together imply G. Gödel seems to suggest that (D) implies (J); I have argued against this in the previous section.

A crucial step in Gödel's argument appears to be an implicit inference from (D) to (G'). The reasoning might be something like this. By (D), when we observe the brain as a physical object, we can distinguish only finitely many states of it, since it is finite. (Finite precision implies that within a finite volume, we can distinguish only finitely many points.) Since the states have to be represented by observably distinguishable brain states, the brain, observing itself 'from inside', can have no special advantage. Otherwise the brain would be able to distinguish more states than is allowed by the finite limit of precision. Is this argument all right? Does (D) imply (G')?

Gödel appears to say that (G) implies (H), which implies (I). Does (G) imply (H)? Does (H) imply (I)? He assigns a higher probability to (D) than (B), although he seems to say also that (D) implies (G'), which implies (B). Does (D) imply (B)?

I should like to conclude this essay by mentioning two other stimulating suggestions by Gödel. In a letter to John von Neumann, dated 20 March 1956, Gödel conjectured that there may be a fast algorithm for deciding, for each mathematical proposition p and each integer n, whether p has a proof of length less than n. (By the way, the problem turns out to be, in today's terminology, an 'NP complete' one.) If the conjecture is true, then, he says, 'one could completely (except for setting up the axioms) replace the mental work of the mathematician concerning yes-or-no problems'.

On 5 June 1976 Gödel told me about a conjecture of his in favour of the power of mind over computers:

It would be a result of great interest to prove that the shortest decision procedure requires a long time to decide comparatively short propositions. More specifically, it may be possible to prove: For every decidable system and every decision procedure for it, there exists some formula of length less than 200 whose shortest proof is longer than 10^{20}. Such a result would actually mean that machines can't replace the human mind, which can give short proofs by giving a new idea.

The conjecture could be seen as a sort of feasible incompleteness theorem. Related to the task of trying to prove something like this, there is the problem of finding an appropriately explicit and interesting formulation of it. For example, it is certainly desirable to state the conditions that a decidable system should satisfy in order that the conjecture holds for it. Also, it may happen that the illustrative formulation, using the specific numbers 200 and 10^{20}, admits counterexamples of an interesting sort so that it would be preferable to use other explicit or not so explicit bounds.

REFERENCES

Ashby, W.R. (1960). *Design for a brain* (2nd edn). Wiley, New York.
Bergson, H. (1946). *The creative mind* (trans. M.L. Andison). Philosophical Library, New York.
Churchland, P.S. and Sejnowski, T. (1992). *The computational brain.* MIT Press, Boston, Mass.
Davis, M. (1965). *The undecidable.* Raven Press, Hewlett, New York.
Edelman, G.M. (1992). *Bright air, brilliant fire: on the matter of the mind.* Basic Books, New York.
Einstein, A. (1949). Reply to criticisms. In *Albert Einstein: philosopher–scientist*, (ed. P.A. Schilpp), pp. 665–88. Open Court, La Salle, Illinois.
Edelman, G.M. (1992). *Bright air, brilliant fire.* Basic Books, New York.
Gödel, K. (1986). *Collected works*, vol.1 (ed. S. Feferman and others). Oxford University Press, New York.
Gödel, K. (1990). *Collected works*, vol.2.
Gödel, K. (1994). *Collected works*, vol.3. Forthcoming. This volume includes Gödel's Gibbs lecture of 1951 and 'The modern development of the foundations of mathematics in the light of philosophy' from about 1962.
Husserl, E. (1970). *The crisis of European sciences* (trans. David Carr). Northwestern University Press.
James, W. (1960). *On psychical research* (compiled and edited by G. Murphy and R.O. Ballou). Viking Press, New York.
Minsky, M. (1988). *The society of mind.* Simon and Schuster, New York.
Penrose, R. (1989). *The Emperor's new mind.* Oxford University Press.
Pour-El, M.P. and Richards, I. (1983). Noncomputability in analysis and physics. *Advances in mathematics*, **48**, 44–74.
Rhees, R. (ed.) (1984). *Recollections of Wittgenstein.* Oxford University Press.
Searle, J. (1992). *The rediscovery of the mind.* MIT Press, Boston, Mass.
Turing, A.M. (1937). On computable numbers. *Proceedings of the London Mathematical Society*, **42**, 230–65; reprinted in Davis (1965), pp. 116–54.
Turing, A.M. (1939). Systems of logic based on ordinals. *Proceedings of the London Mathematical Society*, **45**, 161–228; reprinted in Davis (1965), pp. 155–222.
Turing, A.M. (1950). Computing machinery and intelligence. *Mind*, **59**, 433–60. Reprinted in *The world of mathematics* (ed. J.R. Newman), pp. 2099–123. Simon and Schuster, New York (1956).
Turing, A.M. (1959). Intelligent machinery—a heretical theory. In *Alan M. Turing*, pp. 128–34. Sara Turing, Heffer, Cambridge.
Turing, A.M. (1969). Intelligent machinery. In *Machine intelligence* (ed. B. Meltzer and D. Michie), vol. 5, pp. 1–24. Edinburgh University Press.
Turing, A.M. (1986). Lecture to the London Mathematical Society on 20 February

1947. In *A.M. Turing's ACE Report of 1946* (ed. B.E. Carpenter and R.W. Doran), pp. 106–24. MIT Press, Boston, Mass.

Wang, H. (1974). *From mathematics to philosophy*. Routledge and Kegan Paul, London.

Wittgenstein, L. (1953). *Philosophical investigations* (trans. G.E.M. Anscombe). Basil Blackwell, Oxford.

Wittgenstein, L. (1967). *Zettel* (trans. G.E.M. Anscombe). Basil Blackwell, Oxford.

CHAPTER TWELVE

Knowledge representation and myth

W. F. CLOCKSIN

INTRODUCTION

Most research in artificial intelligence (AI) and cognitive science is founded on the basic premises that intelligence is realized through a representational system based on a physical symbol system, and that intelligent activity is goal-oriented, symbolic activity. The best known expositors of these premises have been Newell and Simon (1972). The essential ingredient is *representation* (the idea that relevant aspects of the world need to be encoded in symbolic form in order to be usable), and research on the practical principles of representation has been a lively area for at least the past twenty years (e.g. Bobrow and Collins 1975). By now most AI and cognitive theories are representationist: the idea that cognition involves the processing of internal 'mental' tokens that stand for external entities or processes. The state of the art is summed up by Hunt (1989) as follows:

The cognitive science movement has been dominated by the *computational view* of thought, which sees thinking as the manipulation of an internal representation ('mental model') of an external domain. The representation must be expressed in some internal language containing designs for well-formed structures and operations upon them. The analogy is more to a computer programming language than to a natural language. Following Fodor (1975), the internal language will be called 'mentalese'. A computational theory of thought must define the mentalese language and describe a hypothetical machine that can execute programs written in it.

In this contribution I shall attempt to begin to show that representationism can be criticized: not everybody feels like taking representationism as the starting-point for cognitive science and AI. Arguments put forward by Maze

(1991) together with experimentation by Brooks and others show that cognitive science can have useful applications if freed from representationist assumptions such as knowledge representation and goal-directed (purposive) behaviour.

I shall also propose a link between cognition and the 'myth-formation process'. The purpose of this is to consider the possibility that the problems of representation cannot be decoupled from the language of myth, and therefore that knowledge representations as expressed by researchers are in fact a form of myth.

The popular conception of myth is associated with the primitive, archaic, and irrational. Myths are considered to be concerned with things so foreign to experience that they cannot be true. But symbol and myth constitute a legitimate way of expressing the transcendent meaning and structure of the knowledge and concerns of all human beings. I suggest that as humans we are so predisposed to narrative expression in the language of myth, that certain conceptually bound endeavours such as the study of knowledge representation cannot be separated from the myth-creation process. Such a predisposition might have a significant effect on the form and meaning of our theories and implementations of knowledge representations.

PROBLEMS OF KNOWLEDGE REPRESENTATIONS

I shall briefly mention three related problems concerning the ways in which the status of knowledge representations can be confused. The best known problem is the 'wishful mnemonic problem' (McDermott 1976). We may attribute intelligence to an AI program because the symbols it manipulates mean something to us, not because what it does with its symbols resembles human intelligence. This problem was originally cast in the form of a pitfall for the unwary programmer, but it applies more generally to the problem of any attribution of meaning to symbols.

A related problem is the 'symbol-grounding problem' (Harnad 1990), in which it is suggested that knowledge representations are not grounded in reality. For example, the predicate formula *holding* (*robot*, *bananas*) does not stand in any relationship to the world. It is we who stand in relation to it. Its meaning is determined by our perceptual world and past experiences.

The third problem is what I call the 'hidden agenda problem': that our methodological assumptions lend an additional and unintended interpretation to the knowledge representations used by a program. For example, if a representationist theory takes as the aim of knowledge representation to 'identify conceptual structures at the level of cognition', this aim has the hidden effect of

(a) building in a structuralist hermeneutic: the idea that the world must be interpreted in terms of an underlying structure that explains, interconnects, and organizes what is perceived;

(b) being self-defining as to scope: the organization of cognition into levels is supplied by the theory, not by the experience of the program;

(c) being self-defining as to detailed nature: representationist theories are founded on introspection or 'interior reflection'. Any type of reflection on the interior life is bound to be expressed in symbolic form.

It could be that the sterility of representationist programs as observed by Brooks (1991) and Winograd (1990) is the result of hidden (and rather limited) agendas being unintentionally built into programs.

It is possible that all these problems are engendered by AI research's affinity with symbolic representations defined within a formal system having classical truth-theoretic semantics or some nonstandard or modal semantics. Two books by Turner (1984, 1990) give a useful state-of-the-art summary of this symbolic logic approach to knowledge representation. But I suggest that this approach is misleading if we expect it to tell us anything about human cognition. Real human cognition and behaviour is not 'rational'—that is, it cannot guarantee that contradictions will not be derived.

The fact that some people, namely trained logicians, might be rational on paper for a few moments each day does not refute the point. The performance of any type of derivation or calculation is a symbol game, the rules of which can be applied with more or less precision by a suitably trained person. The development of logic is useful to understand argument and for computer programming by reducing these to a symbol game, but using it to characterize knowledge representations in the service of cognition and behaviour is a job for which it was never intended, and for which a number of researchers consider it unsuited.

EXPERIMENTS WITH REPRESENTATION-FREE SYSTEMS

Recent research has been involved with carrying out complicated perception and action tasks without explicit representations and without the requirement for articulating the goal structure of a problem. It may be argued that the general direction of connectionist research is toward representation-free systems, particularly when the aim is to train networks to control systems whose input/output relationship is too difficult or impractical or impossible to model analytically. However, this aim is not exclusive to connectionism, and connectionism has not stopped researchers from trying to build representationist schemes into networks (Hinton 1987). The value of connectionist research is not that it has proved the value of neural networks (it has not), but that it has motivated investigators who might otherwise not do so to consider 'black box' techniques for developing systems that construct their own input/output relations. Apart from this, most results in connectionism may be placed in the category of implementation techniques.

One example of the investigation of representation-free systems is the State-Space Robotics project (Clocksin and Moore 1989). The experimental system consists of a six-degrees-of-freedom robot arm and two video cameras connected to a minicomputer. The system engages in unattended unsupervised real-time trial-and-error learning of reaching and rotary pursuit tasks, and after a short period of learning, the system carries out the tasks successfully. The behaviour of the system is such that it appears as though the system designers have 'solved' the so-called correspondence problem of binocular vision and have 'solved' the so-called real-time inverse dynamics equations for the robot arm. In fact, no such algorithms or explicit representations were used. The system operates by trial-and-error filling-in of a ten-dimensional state-space memory that represents the product space of each perceived variable (derived from the video signal) and each output variable (signal to robot arm joints). This space represents the coordination relation of the robot situated in its environment. The robot behaviour incrementally converges toward behaviour that is 'naturally-selected' by the environment, and which is not aided by any sort of built-in trajectory plan. The state-space memory is implemented using the textbook technique of a k-D binary tree. The learning procedure, which involves finding near-neighbours in the tree, converges quickly without the requirement for repeated presentation of data. The tree-like structures that form in the memory as a result of learning are at least as resemblant of biological neural structures as are the so-called neural networks.

Situated Robotics (Brooks), State-Space Robotics (Clocksin and Moore), and Animate Perception (Ballard 1991; Whitehead and Ballard 1991) are three of the several independent strands of research that turn out to have a common aim: that representations are not built into the systems by a designer. Any *post hoc* analysis of a system's memory contents may reveal structures which—to a third-party observer—appear to be representational, but the point is that the system's designer did not put them there. More importantly, the system's designer did not specify rules of formation by which these structures are constructed and modified in a context-dependent way in the course of the system's operation.

A MYTH-HERMENEUTIC OF COGNITION AND REPRESENTATION

The points raised in the following discussion have been inspired by the work of the analytical psychologist C. G. Jung. Some acquaintance with his ideas of the archetype, myth, and the collective unconscious is assumed.

Most AI researchers agree that the aim of knowledge representation is to identify conceptual structures at the level of cognition. Leaving aside for the moment the implicit structuralist hermeneutic of that statement, knowledge representations try to represent an interior life, and they originate with some sort of introspection or interior reflection. Owing to our heritage (from species and culture) of 'interior archaeology' as studied by psychoanalysts, any type of reflection on the interior life is bound to be expressed in symbolic form—in the same language that has become associated with myths.

Myth must therefore, be considered a characteristic of being human, and is not merely a matter of archaic primitivism used only by unintelligent people who can understand in no other way. Nor is myth the sole property of peoples whose technology has not developed (indeed, the advanced technology of industrialized countries is such as to project and empower our myths to a high degree). Nor is myth a matter of the decay of reason or of a disordered unconscious. I shall not discuss those aspects of myth usually of interest to anthropologists, in which myth is merely a practical service to answer questions of aetiology and eschatology: for personal empowerment, to justify existing social systems, and to account for traditional rites and customs.

In the devising and use of knowledge representations as technical artefacts, we are dealing with a myth-creation process. Researchers often think of knowledge representations as animated, metaphorically alive. The representations 'live' in a space of human concerns and have components that store data of interest to us. They communicate among themselves along links (or connections or channels) and even reproduce (the 'spawning' of processes is a frequently used metaphor).

During the past thirty years, artificial intelligence research has unintentionally given rise to a whole new mythology. There is a level at which aspects of standard knowledge-representation techniques can be identified with the archetypes of the collective unconscious. For example, the roles that nodes may take as suppliers (*animus*) or receivers (*anima*) of data, the organization into layers or hierarchy, and the minor pantheon of active and autonomous 'agents'. The 'binding' and 'substitution' of variables within structures corresponds directly to notions of substitution in the semiotics of myth. There are many more such correspondences to be found by those disposed to look for them.

The use of formal logic as a framework for representing knowledge can be seen not only as an operational technique, but as a particularly rich projection of archetypes. The methodology of rigorous axiomatizing of knowledge—which flourishes among AI research notwithstanding the fact that human 'rationality' is frail and fallible—may on the one hand be identified with a compulsion for order, deeply rooted in the unconscious as the masterly 'hero' archetype. Alternatively, the adherence to logic as a representational framework can be seen also as a projection of the archetype of the virgin: sublime, immune to the pitfalls of human weakness, and free from the defects of the world. Finally, the idea that knowledge representation in the service of cognition proceeds by means of the matching of terms according to the unification algorithm (or similar algorithms) has an essential archetypal meaning in terms of the union of the 'above' (the term in the AI program) with the 'below' (the term in the input data): in archetypal terms this is associated with the dream of the earth that reaches up to touch the sky; the symbol of ultimate unity in the mythic concept of a divine marriage. The fruit of this marriage is the binding of variables to terms; so the logician's concept of instantiation is archetypally related to the theologian's concept of incarnation.

Researchers often ask each other whether there is a psychological validity to their work on knowledge representation. Certainly the psychocultural/ psychoanalytic validity seems to be there already in the form that representations have taken, but I expect this comes as little comfort, for it says more about the researchers themselves than about the validation of particular scientific results.

PURPOSIVE BEHAVIOUR

One example of a myth is the notion of goal-directed (purposive) behaviour. In particular, researchers' desires to attribute concepts such as purpose to animals need explanation. People seem to have a predilection for teleological explanations of behaviour. For example, some researchers of animal behaviour have observed what appears to be cheating, lying, and other forms of deceit among populations of (non-human) animals. This is taken as evidence for the idea that these animals manipulate and represent knowledge, since it is assumed that a situation needs to be represented to a sufficient degree before the animal can decide to use the representation as information necessary to plan an act of deceit. But according to the myth-hermeneutic, the idea that the animal represents knowledge is actually a myth, constructed not by the animal, but by the researcher, who according to psychoanalytic theory is drawing upon the primitive 'trickster' archetype in his unconscious in order to interpret his observations of the animal.

There is no reason at all to believe that an animal's act that appears to us to be deceitful is intentional, even if the behaviour is reliably reproducible. For, what appear to be goals may well be achieved as a consequence of naturally selected behaviour without supposing that the animal 'had in mind' what it 'wanted' to achieve. There is no known limit to the complexity of a 'mere' reflex behaviour, even without considering the evolution of such behaviour over the time span of a whole species in its environment. Even the most basic of finite-state machines can be programmed to exhibit complicated forms of deceit without an inbuilt knowledge-representation scheme. The fact that a suitable program need not be written by a person, but may simply evolve over a period of time as the result of a successive enumeration of incrementally perturbed programs, is in addition to but beside the point.

So people seem to be predisposed to assume that deceitful acts by animals are intentional. Such assumptions are not restricted to animals and deceitful acts. The experiments of Michotte showed vividly that people are willing to attribute not only intentions but whole personalities to little coloured cardboard squares moving about on a screen. We all know that cardboard squares do not have self-awareness, and the attribution of personalities is just an entertaining fantasy. As for machines, we think that machines do not have self-awareness, but many people like to think it might be possible to program machines to have self-awareness. Equipping machines with what we consider to be the necessary mechanism for intelligent behaviour is not considered a fantasy, but a serious business for many AI researchers. As for animals, we are genuinely not sure of the extent of self-awareness, so it is easier for us to attribute ratiocination and its presumed attendant knowledge representations to animals. As for people, we take such behaviour for granted, and furthermore we seem to be generous in the attribution of human characteristics to objects and creatures other than humans. The point is that according to the mythic-hermeneutic, all these attributions have a common cause and cognitive foundation: the myth-formation process at work in the most primitive layers of our psyche.

DISCUSSION

Knowledge representations are outward expressions of an interior life, and the act of investigating knowledge representations is also an outward expression of the interior life. Thus it should not be surprising that archetypes of the unconscious are stirred to conscious expression by any encounter with computational realities that might correspond to their meaning. We can therefore expect that artificial intelligence research, particularly in the area of knowledge representation, should be a great call to the forces of the unconscious, because AI has the aim of investigating cognition itself via what are presumed to be working models. Yet, AI researchers are not accustomed to the methodology of discernment: in this case the querying of the extent to which their new knowledge representation technique or language really is an objective representation of some aspect of cognition, or the extent to which it is a manifestation of 'the images that surge and tumble in the unconscious archaeological layers of our psyche', to use Boff's (1979) phrase.

Some philosophers of science have also suggested that science and myth are closer relations than has been previously admitted. For example, Hesse (1983) suggests that it is a mistake to interpret the pressure towards universalizable science as the search for a comprehensive true theory corresponding to reality. Instead, scientific theory is interpreted as one type of response to cultural needs for myth and ideology. Thus, conflicting mythologies (including scientific ones) should not be seen as autonomous atoms within the social milieu, but as interacting systems of cognition and value. Popper (1992) has commented that 'poetry and science have the same origin. They originate in myths.'

What form might a cognitive architecture based on a mythic-hermeneutic take? First, the assumption is that at the lowest level of cognition there is found a number of primitive archetypes that interact with perceptions and memories to form subconscious patterns of activation I shall call subnarratives. After further elaboration and interaction with perception and memory, some of these subnarratives may surface into a conscious life to be articulated as myths. These myths probably take a primitive form of narratives which are articulated over time and are closely coupled with rhythmic motor behaviour. Such a primitive form is to be contrasted to the general AI view of symbolic conceptual statements or a network of such statements that 'reside' in the brain.

It is possible that further insight may be obtained from Cupitt's (1991) work on narratives, although from Cupitt's standpoint he must consider narrative to fulfil a cultural need rather than the basic organic need proposed here. I regard cultural influences as essential to the development of cognition, but they function to constrain the variety of perceptions and memories that are exposed to the system, and consequently to constrain the range of possible narratives generated by the system.

Precursors of the architecture may also be discerned from Minsky's (1985) concept of the society of mind, in which multiple mental agents are specialized in primitive tasks and are grouped in ways to carry out more complicated tasks. Minsky's Builder and Wrecker agents, for example, may relate to primitive archetypes. Minsky's description of Wrecker strongly suggests the Jungian Trickster archetype, while Builder is probably a composite of more primitive Jungian archetypes in which the Hero features. Minsky's notion of frames can be seen as a third-party description of how narratives might be assembled. This of course does not necessarily imply that frames have any further instrumental or representational significance. I consider narratives to be concept-free, contextualized only within the behaviour of the neural substrate.

CONCLUSION

There is a connection between symbol, myth, and an important area of AI research: the representation of what are presumed to be cognitive processes underlying knowledge and behaviour. I have suggested how knowledge representation might be interpreted according to a hermeneutic of myth. Any practical consequence of this approach will involve the implementation of a cognitive architecture based on archetypes and narratives, in contrast to the more widely known architecture based on, for example, the knowledge base and inference engine with its essential requirement for logical representation. And perhaps one day Marvin Minsky's book *The society of mind* will be affectionately remembered, not as a work of science, but as one of the foremost mythologies of our time.

REFERENCES

Ballard, D.H. (1991). Animate vision. *Artificial intelligence*, **48**, 57–86.

Bobrow, D. and Collins, A. (eds) (1975). *Representation and understanding.* Academic Press, London.

Boff, L. (1979). *The maternal face of God.* Collins, London.

Brooks, R.A. (1991). Intelligence without representation. *Artificial intelligence*, **47**, 139–60.

Clocksin, W.F. and Moore, A.W. (1989). Experiments in adaptive state-space robotics. In *Proceedings of the seventh conference of the society for artificial intelligence and simulation of behaviour* (ed. A.G. Cohn), pp. 115–25.

Cupitt, D. (1991). *What is a story?* SCM Press, London.

Fodor, J. (1975). *The language of thought.* Harvester Press, Hemel Hempstead, Herts.

Harnad, S. (1990). The symbol grounding problem. *Physica D*, **42**, 335–46.

Hesse, M. (1983). Cosmology as myth. *Concilium*, **166**, 49–54.

Hinton, G.E. (1987). Learning distributed representations of concepts. *Proceedings of the cognitive science society* (CSS-9), 8.

Hunt, E.B. (1989). Cognitive science: definition, status and questions. *Annual Review of Psychology*, **40**, 603–29.

McDermott, D.V. (1976). Artificial intelligence meets natural stupidity. *SIGART Newsletter*, **57**, 4–9. Reprinted in *Mind design* (ed. J. Haugeland), pp. 143–60. MIT Press, Cambridge, Mass.

Maze, J.R. (1991). Representationism, realism and the redundancy of 'mentalese'. *Theory and Psychology*, **1**(2), 163–85.

Michotte, A. (1963). *The perception of causality*. Basic Books, New York.

Minsky, M. (1985). *The society of mind*. Simon and Schuster, New York.

Newell, A. and Simon, H.A. (1972). *Human problem solving*. Prentice-Hall, Englewood Cliffs, NJ.

Popper, K. (1992). *In search of a better world: lectures and essays from thirty years*. Routledge, London.

Turner, R. (1984). *Logics for artificial intelligence*. Ellis Horwood, Chichester, Sussex.

Turner, R. (1990). *Truth and modality for knowledge representation*. Ellis Horwood, Chichester, Sussex.

Whitehead, S.D. and Ballard, D.H. (1991). Learning to perceive and act by trial and error. *Machine learning*, **7**, 45–83.

Winograd, T. (1990). Thinking machines? Can there be? Are we? In *The foundations of artificial intelligence* (ed. D. Partridge and Y. Wilks). Cambridge University Press.

CHAPTER THIRTEEN

Memory and the individual soul: against silly reductionism

GERALD M. EDELMAN

From the last quarter of the seventeenth century to the last decade of the eighteenth, an explosion of creativity called the Enlightenment changed the history of ideas. Its reigning views were many, but above all it was dedicated to reason, to science, and to human freedom and individuality. Its underlying science was physics, the system of Newton, and its philosophy of society was, in large measure, that of Locke. Yet the Enlightenment ideas of causality and determinism, together with its mechanistic view of science, undermined hopes for a theory of human action based on freedom. If we are determined by natural forces—by mechanism—we cannot easily put together a consistent picture in which a free individual makes moral choices. Moreover, while the ideas of the Enlightenment paid much attention to the role of reason and culture in such choices, there was no general notion of how deeply the minds of all humans (including those of 'reasonable' human beings—that is, the 'cultured') were influenced by unconscious forces and by emotion.

Whatever forms it took at various times and places, the overriding Enlightenment view was a secular one that forged many of the ideas underlying modern democracy. But despite its valuable heritage, the Enlightenment is over. The first great blow to its ideas came with Hume's damaging attacks on both rationalism and the notion of human progress as linked to natural science. Its major fault was its inability to create an adequate scientific description of a human individual to accompany its description of a machine-like universe. Its social failure was its inability to go beyond the concept of a society composed of self-seeking, commercially successful individuals with a shallow view of 'humanism'. Certainly, Enlightenment thinkers attempted to provide us with a larger, more inspiring view of ourselves. But its science was a mechanistic physics and it had no body of data or ideas with which to link the world, the mind, and society in the style of scientific reason to which it aspired. Whatever the Enlightenment's failures and inconsistencies, however, it left us with high hopes for the place of the individual in society.

Can we expect to do better with a sound scientific view of mind? *I hope to show that the kind of reductionism that doomed the thinkers of the Enlightenment is confuted by evidence that has emerged both from modern neuroscience and from modern physics. I have argued that a person is not explainable in molecular, field-theoretical, or physiological terms alone.* To reduce a theory of an individual's behaviour to a theory of molecular interactions is simply silly, a point made clear when one considers how many different levels of physical, biological, and social interactions must be put into place before higher-order consciousness emerges. The brain is made up of 10^{11} cells with at least 10^{15} connections. Each cell has a fantastically intricate regulatory biochemistry constrained by particular sets of genes. These cells come together during morphogenesis and exchange signals in a place-dependent fashion to make a body and a brain with enormous numbers of control loops, all obeying the homeostatic mechanisms that govern survival. Selection on neuronal repertoires leads to changes in myriad synapses as cells die or differentiate. An animal's survival and motion in the world allow perceptual and conceptual categorization to occur continually in global mappings. *Memory dynamically interacts with perceptual categorization by re-entry.* Learning involving the connection of categorization to value (in its most subtle form within a speech community) links symbolic and semantic abilities to conceptual centres that already provide embodied structures for the building of meaning.

A calculation of the significant molecular combinations of such a sequence of events, even in identical twins, is almost impossible and, in any case, useless. The mappings are many—many, and the processes are individual and irreversible. I wonder what Enlightenment humanists would have made of all this. Diderot, who speculated about the nervous system of his friend in *Le Reve de d'Alembert*, might have been pleased. Diderot's view of human consciousness opened up the possibility that to be human was to go beyond mere physics.

I have taken the position that there can be no complete science, and certainly no science of human beings, until consciousness is explained in biological terms. Given our view of higher-order consciousness, this also means an account that explains the bases of how we attain personhood or selfhood. By selfhood I mean not just the individuality that emerges from genetics or immunology, but the personal individuality that emerges from developmental and social interactions.

Selfhood is of critical philosophical importance. Some of the problems related to it may be sharpened by the selectionist view I have taken on the matter of mind. Please remember, however, that no scientific theory of an individual self can be given (our qualia assumption). None the less, I believe that we can progress toward a more complete notion of the free individual, a notion that is essential to any philosophical theory concerned with human values.

The issues I want to deal with are concerned with the relationship between consciousness and time, with the individual and the historical aspects of memory, and with whether our view of the thinking conscious subject alters our notion of causality. I also want to discuss briefly the connection between emotions and our ideas of embodied meaning. All these issues ultimately bear upon the matter of free will and therefore upon morality under mortal conditions.

According to the extended Theory of Natural Group Selection (TNGS), memory is the key element in consciousness, which is bound up with continuity and different time-scales. There is a definite temporal element in perceptual categorization, and a more extended one in setting up a conceptually based memory. The physical movements of an animal drive its perceptual categorization, and the creation of its long-term memory depends on temporal transactions in its hippocampus. As we have seen, the Jamesian properties of consciousness may be derived from the workings of such elements. But in human beings, primary consciousness and higher-order consciousness coexist, and they each have different relations to time. The sense of time past in higher-order consciousness is a *conceptual* matter, having to do with previous orderings of categories in relation to an immediate present driven by primary consciousness. Higher-order consciousness is based, not on ongoing experience, as is primary consciousness, but on the ability to model the past and the future. At whatever scale, the sense of time is first and foremost a conscious event.

The ideas of consciousness and 'experienced' time are therefore closely intertwined. It is revealing to compare the definition of William James, who stated that consciousness is something the meaning of which 'we know as long as no one asks us to define it', with the reflections of St Augustine, who wrote in his *Confessions*, 'What then is time? If no one asks me, I know what it is. If I wish to explain to him who asks me, I do not know.' The notion of continuity in personal, historical, and institutional time was a central one in Augustine's thought.

Time involves succession. An intriguing suggestion about the connection between time and the idea of numbers has come from L. E. J. Brouwer, a proponent of intuitionism in mathematics. He suggests that all mathematical elements (and particularly the sequence of natural numbers) come from what he calls 'two-icity'. Two-icity is the contrast between ongoing conscious experience (with primary consciousness as a large element) and the direct awareness of post experience (requiring higher-order consciousness). What is intriguing about this is that it suggests that one's concept of a number may arise not simply from perceiving sets of things in the outside world. It may also come from inside—from the intuition of two-ness or two-icity plus continuity. By recursion, one may come to the notion of natural numbers.

Whatever the origins of such abstractions, the personal sense of the sacred, the sense of mystery, and the sense of ordering and continuity all have connections to temporal continuity as we experience it. We experience it as individuals, each in a somewhat different way.

Indeed, the flux of categorization, whether in primary or higher-order consciousness, is an individual and irreversible one. It is a history. Memory grows in one direction; with verbal means, the sense of duration is yet another form of categorization. This view of time is distinguishable from the relativistic notion of clock time used by physicists, which is, in the microscopic sense, reversible. Aside from the variation and irreversibility of *macroscopic* physical events recognized by physicists, a deep reason for the irreversibility of individually experienced time lies in the nature of selective systems. In such systems, the emergence of pattern is *ex post facto*. Given the diversity of the repertoires of the brain, it is extremely unlikely that any two selective events, even apparently identical ones, would have identical consequences. Each individual is not only subject, like all material systems, to the second law of thermodynamics, but also to a multilayered set of irreversible selectional events in his or her perception and memory. Indeed, selective systems are by their nature irreversible.

This 'double exposure' of a person—to real-world alterations affecting nonintentional objects as well as to individual historical alterations in his or her memory as an intentional subject—has important consequences. The flux of categorizations in a selective system leading to memory and consciousness alters the ordinary relations of causation as described by physicists. A person, like a thing, exists on a world-line in four-dimensional space–time. But because individual human beings have intentionality, memory, and consciousness, they can sample patterns at one point on that line and on the basis of their personal histories subject them to plans at other points on that world-line. They can then enact these plans, altering the causal relations of objects in a definite way according to the structures of their memories. It is as if one piece of space–time could slip and map onto another piece. The difference, of course, is that the entire transaction does not involve any unusual piece of physics, but simply the ability to categorize, memorize, and form plans according to a conceptual model. Such a historical alteration of causal chains could not occur in so rich a way in any combination of inanimate nonintentional objects, for they lack the appropriate kind of memory. This is an important point in discriminating biology from physics.

In certain memorial systems, unique historical events at one scale have causal significance at a very different scale. If the sequence of an ancient ancestor's genetic code was altered as a result of that ancestor's travels through a swamp (driven, say, by climatic fluctuations), the altered order of nucleotides, if it contributed to fitness, could influence present-day selectional events and animal function. Yet the physical laws governing the actual *chemical* interaction of the genetic elements making up the code (the nucleotides) are deterministic. No deterministic laws at the chemical level could alone, however, explain the *sustained* code change that was initiated and then stabilized over long periods as a result of complex selectional events on whole animals in unique environments.

Memorial events in brains undergoing selectional events are of the same ilk. Because the environment being categorized is full of novelty, because selection is *ex post facto*, and because selection occurs on richly varied historical repertoires in which different structures can produce the same result, many degrees of freedom exist. We may safely conclude that, in a multilevel conscious system, there are even greater degrees of freedom. These observations argue that, for systems that categorize in the manner that brains do, there is macroscopic indeterminancy. Moreover, given our previous arguments about the effects of memory on causality, consciousness permits 'time slippage' with planning, and this changes how events come into being.

Even given the success of reductionism in physics, chemistry, and molecular biology, it nonetheless becomes silly reductionism when it is applied exclusively to the matter of the mind. The workings of the mind go beyond Newtonian causation. The workings of higher-order memories go beyond the description of temporal succession in physics. Finally, individual selfhood in society is to some extent a historical accident.

These conclusions bear on the classical riddle of free will and the notion of 'soft determinism', or compatibilism, as it was called by James Mill. If what I have said is correct, a human being has a degree of free will. That freedom is not radical, however, and it is curtailed by a number of internal and external events and constraints. This view does not deny the influence of the unconscious on behaviour, nor does it underestimate how small biochemical changes or early events can critically shape an individual's development. But it does claim that the strong psychological determinism proposed by Freud does not hold. At the very least, our freedom is in our grammar.

These reflections, and the relationship of our model of consciousness to evolved values bear also on our notion of meaning. Meaning takes shape in terms of concepts that depend on categorizations based on value. It grows with the history of remembered body sensations and mental images. The mixture of events is individual and, in large measure, unpredictable. When, in society, linguistic and semantic capabilities arise and sentences involving metaphor are linked to thought, the capability to create new models of the world grows at an explosive rate. But one must remember that, because of its linkage to value and to the concept of self, this system of meaning is almost never free of affect; it is charged with emotions. This is not the place to discuss emotions, the most complex of mental objects, nor can I dedicate much space to thinking itself. But it is useful to mention them here in connection with our discussion of free will and meaning. As philosophers and psychologists have often remarked, the range of human freedom is restricted by the inability of an individual to separate the consequences of thought and emotion.

Human individuals, created through a most improbable sequence of events and severely constrained by their history and morphology, can still indulge in extraordinary imaginative freedom. They are obviously of a different order from nonintentional objects. They are able to refer to the world in a variety of ways. They may imagine plans, propose hopes for the future, and causally affect world events by choice. They are linked in many ways, accidental and otherwise, to their parents, their society, and the past. They possess 'selfhood', shored up by emotions and higher-order consciousness. And they are tragic, in so far as they can imagine their own extinction.

It is often said that modern humans have suffered irreversible losses from several episodes of decentration, beginning with the destruction of earlier cosmologies placing human beings at the centre of the universe. The first episode, according to Freud, however, took place when geocentrism was displaced by heliocentrism. The second was when Darwin pointed out the descent of human beings. And the third occurred when the unconscious was shown to have powerful effects on behaviour. Well before Darwin and Freud, however, the vision of a Newtonian universe led to a severe fatalism, a view crippling to the societal hopes of Enlightenment thought. Yet we can now see that if new ideas of brain function and consciousness are correct, this fatalistic view is not necessarily justified. The present is not pregnant with a fixed programmed future, and the programme is not in our heads. The theories of modern physics and the findings of neuroscience rule out not only a machine model of the world but also such a model of the brain.

We may well hope that if sufficiently general ideas synthesizing the discoveries that emerge from neuroscience are put forth, they may contribute to a second Enlightenment. If such a second coming occurs, its major scientific underpinning will be neuroscience, not physics.

The problem then will be not the existence of souls, for it is clear that each individual person is like no other and is not a machine. The problem will be to accept that individual minds are mortal. Given the secular views of our time, inherited from the first Enlightenment, how can we maintain morality under mortal conditions? Under present machine models of the mind this is a problem of major proportions, for under such models it is easy to reject a human being or to exploit a person as simply another machine. Mechanism now lives next to fanaticism: societies are in the hands either of the commercially powerful but spiritually empty or, to a lesser extent, in the hands of fanatical zealots under the sway of unscientific myths and emotion. Perhaps when we understand and accept a scientific view of how our mind emerges in the world, a richer view of our nature and more lenient myths will serve us.

How would humankind be affected by beliefs in a brain-based view of how we perceive and are made aware? What would be the result of accepting the ideas that each individual's 'spirit' is truly embodied; that it is precious *because* it is mortal and unpredictable in its creativity; that we must take a sceptical view of how much we can know; that understanding the psychic development of the young is crucial; that imagination and tolerance are linked; that we are at least all brothers and sisters at the level of evolutionary values; that while moral problems are universal, individual instances are necessarily solved, if at all, only by taking local history into account? Can a persuasive morality be established under mortal conditions? This is one of the largest challenges of our time.

What will remain unclear until neuroscience grows more mature is how any of these issues can be linked to our history as individuals in a still-evolving species. In any case, silly reductionism and simple mechanism are out. A theory of action based on the notion of human freedom—just what was missing in the days of the Enlightenment—appears to be receiving more and more support from the scientific facts.

INDEX

activation vectors 73
adaptive robots/systems 151, 153
aether, elimination from serious science 69, 70
age of Universe 54
agnosia 114
Ahlfors, Lars 3
AI, *see* artificial intelligence
algorithmically incompressible/irreducible
    information 34, 36
algorithmic information theory
    early version 34
    with self-delimiting programs 34
algorithmic nature of physics 171, 174–9
A-life 153
    see also artificial intelligence
amnesia effects 113–14
anaesthesia, effects 112
anosognosia 113
anti-clericalism 142
ants, as models of automata 150
Aristotelian approach 51
arithmetic, randomness in 37–40
art form, science as 11
artificial intelligence (AI)
    and ants 150–4
    and biology 150, 154
    conscious thought modelled by 157–8
    and human dignity 148–59
Ashby, W. Ross 172
Aspect, Alain 17
astronomy, experiments 46
atheism 128, 142, 143
Atkins, Peter W. 122–32, 137, 138
Augustinian view of time 202
austerity [of reductionism] 134–5
    pay-offs resulting 134, 143
automata
    Darwin series 92–8, 117
    early designs 92–3, 94, 151
    see also robots
autonomous agents 154
autonomy 154–8
    environment response criterion 154, 155
    reflexively modifiable directing mechanisms
        criterion 154, 155, 157–8
    self-generated control criterion 154, 155–7

Barash, David, quoted 145
Barrow, John D. 45–62
behaviourism 143, 149
    depersonalizing effects 143, 150
Bell, John 17, 178
Bell nonlocality 17, 178
Bergson, Henri 188
Bernal, J. D., quoted 135, 146–7
Bernard, Claude 108
Bernstein problem 92
Besso, Michele 137
big-bang model [of Universe] 56
binding problem 91, 112
biology, distinguished from physics 138, 203–4

black holes 6–7, 8
    Einstein's hostility to concept 6–7
    and general relativity theory 8
block-sorting tasks [by recognition automaton]
    97–8
Boden, Margaret A. 148–59
body-image/body-ego disturbances 112–13
Boids [in computer-modelled flocking behaviour]
    153
bootstrapping [in mind] 110, 111
Borel, Emile 35
brain
    as computer 23, 78, 79, 99, 149, 169–74
    development in size explained by TNGS 111
    number of neurones and synapses 165, 201
    as selectional system 78–9, 104–5, 111, 172
    as site of mind 163
    see also nervous systems
brain action
    holism in 21–6
    scientific explanation 130
brain activity, quantum effects 22
brains, effects of failure 130
Brooks, R.A. 191, 192
Brouwer's intuitionist approach to mathematics 43,
    44, 202
Bruner, Jerome 116
Burnet, MacFarlane 104

Cantor's diagonal procedure 30, 31–2
carbon-based life 54
categorization
    irreversibility of 203
    value-based 105, 205
cell adhesion molecules (CAMs) 80, 81
Chaitin, Gregory J. 27–44, 47
Champernowne, David 35
Champernowne's number 35, 36
chaos, order out of 46–7, 127
chaos theory 158
child development theory 114–15
Churchland, Patricia S. 64–76, 171
Churchland, Paul M. 64–76
Church's thesis 170
classical physics
    algorithmic nature of 176
    inability to explain consciousness 23
    relationship to quantum physics 23
classification couples [re-entrantly connected maps]
    81, 83
Clocksin, William F. 190–8
clonal selection [in immunity] 79, 104
cognitive science 116
    representationist basis 190
cohomology 14–15
collagen [triple-helix] structure 10
collective unconscious 194
colour-vision theory 72
combustion theories 69
completeness 27
complexity
    organized structures 52–3